LIVING WATERS

Ecology of Animals in Swamps, Rivers, Lakes and Dams

Nick Romanowski

PUBLISHING

National Library of Australia Cataloguing-in-Publication entry

Romanowski, Nick, 1954– author.
Living waters : ecology of animals in swamps, rivers, lakes and dams / by Nick Romanowski.

9780643107564 (paperback)
9780643107571 (epdf)
9780643107588 (epub)

Includes bibliographical references and index.

Wetlands – Australia.
Wetland ecology – Australia.
Wetland conservation – Australia.
Biodiversity conservation.

333.9180994

Published by
CSIRO PUBLISHING
36 Gardiner Road, Clayton VIC 3168
Private Bag 10, Clayton South VIC 3169
Australia

Telephone: [+613] 9545 8555
Local call: 1300 788 000 (Australia only)
Fax: +61 3 9662 7555
Email: csiropublishing@csiro.au
Website: www.publishing.csiro.au

All images are by Nick Romanowski.

Set in 11/13.5 Minion
Edited by Joy Window
Cover and text design by James Kelly
Typeset by Thomson Digital, Chennai, India
Printed by Ingram Lightning Source

Feb26_RP_ILS

Contents

	Acknowledgements	*vii*
	Introduction	*ix*
Part 1	**The diversity of wetland animals**	**1**
Chapter 1	**Invertebrates: the crustaceans**	**3**
	Copepods	4
	Plants: phytoplankton and zooplankton	6
	Other smaller crustaceans	7
	Crayfishes, shrimps and crabs	12
	Moulting and growth	15
Chapter 2	**Invertebrates: the insects**	**17**
	The insect body plan	17
	Wetland insects: the major groups	21
	Emergence of a dragonfly	23
	Metamorphosis	26
Chapter 3	**Other invertebrate players**	**35**
	Miniscule worlds	35
	Molluscs	37
	The nature of water: O_2, CO_2 and breathing	38
	Worms, leeches, sponges and jellyfish	40
	Spiders and mites	42
	Parasites and other strange life forms	43
Chapter 4	**Fishes**	**47**
	Moving between the sea and fresh waters	48
	Finding your way in murky waters	51
	The diversity of native freshwater fishes	52
	Human impacts: introduced fishes	59

Chapter 5	**Frogs, reptiles and mammals**	**61**
	The life cycles of frogs	62
	Wetland frogs	64
	Freshwater turtles	65
	Lizards, snakes and crocodiles	67
	Human impacts: alien amphibians ... and a reptile	69
	Mammals	71
	Origins: Gondwanaland and Asia	72
Chapter 6	**Waterbirds**	**75**
	The migratory urge	75
	Human impacts: waterbirds, environmental flows and agriculture	78
	Eggs, nests and later problems	80
	The diversity of wetland birds	81
	Australian pelicans	89
	Inland shorebirds and seabirds	92
Part 2	**Living with change**	**95**
Chapter 7	**Making the most of change**	**97**
	Small worlds	97
	Crabs and crayfishes	101
	Plants: paperbarks and water-ribbons	104
	Frogs	105
	Fishes	107
	Birds	109
	Human impacts: drainage	112
Chapter 8	**Moving on**	**115**
	Fishes and the upstream urge	115
	Origins: moving inland from the sea	120
	Overland or by air	121
	Insects	123
	Waterbirds – moulting and migration	125
	Human impacts: climate change	127
Chapter 9	**Rebirth of a lake**	**129**
	Drought and the new millennium	130
	Plants: movement	132
	Returning waters	133
	The feathered hordes	138
	Lake Corangamite	143

Chapter 10	**Predator and prey**	**145**
	What it takes to be a predator	146
	Hidden predators – arthropods	148
	Fishes as predators and prey	152
	Waterbirds as predators	154
	Crocodiles	156
	Human impacts: hunting, fishing and aquaculture	158
Part 3	**Ecology: fitting into your environment**	**161**
Chapter 11	**High and low places**	**163**
	Water flowing underground	163
	Carnivorous plants	166
	Where the salt goes	170
	Troglodytes	175
Chapter 12	**Streams, brooks and fast rivers**	**177**
	On higher ground	178
	Staying put in moving waters	179
	Spiny crayfishes	181
	Fishes and fast streams	183
	Galaxiids: cryptic diversity in the south	186
	Frogs and reptiles	187
Chapter 13	**Slow rivers in an ancient land**	**189**
	Plants: river red gum	191
	Fishes of the inland rivers	192
	Turtles and invertebrates … but few frogs	195
	Waterbirds: drifters and stayers	197
	Human impacts: why the bunyip disappeared	200
Chapter 14	**Floodplains, billabongs and backwaters**	**203**
	Human impacts: weeds	205
	Insects and crustaceans	206
	Fishes, frogs and reptiles	208
	Rainbowfishes: flamboyant diversity in the north	211
	Waterbirds	212
Chapter 15	**Marshes and swamps**	**215**
	Swamp plants: diversity and interactions	215
	Plants: sedges, reeds and rushes	217

	Invertebrate lifestyles	219
	The nature of water: surface tension	221
	Frogs, tadpoles and waterbirds	223
Chapter 16	**Freshwater lakes, lagoons and other dark waters**	**227**
	The nature of water: black waters – peat and tannins	228
	Fishes and frogs	230
	Waterbirds and platypus: guides to hidden worlds	233
Chapter 17	**Estuaries, mangroves and saltmarshes**	**237**
	Plants: saltmarsh and mangroves	239
	Shifting lives in estuaries	241
	Fishes of estuaries and streams	242
	Origins: gobies (and gudgeons)	246
	Reptiles and fishes in the north	248
	Other coastal wetlands	249
Chapter 18	**New opportunities: dams and other created wetlands**	**255**
	What wetland animals need	257
	Attracting waterbirds	258
	Fishes and frogs: finding a balance	261
	Hidden worlds	263
	Created wetlands and habitat	266
	Glossary	267
	Further reading	271
	Guide to common and scientific names used in this book	277
	Index	285

Acknowledgements

It is now more than 50 years since I first started learning about the varied animals, plants and ecosystems that make up the rich tapestry of Australian wetlands, stretching from cold temperate climates whipped by blizzards to the monsoonal tropics, and extending far inland into some of the driest habitable places on Earth. I was fortunate to have these early interests encouraged and organised at Monash University, when Ian Bayly and David Williams were creating a renaissance in the study of the ecology of Australian inland waters.

The process was accelerated a few years later when I set up Australia's first indigenous wetland nursery, *Dragonfly Aquatics*, originally a means of financing my freelance studies of plant and animal interactions, though it rapidly grew to become a hands-on introduction to the ecology of the plants themselves. Since that time I have been fortunate to have been tutored by many people with diverse interests, backgrounds and expertise relating to wetlands, but for those formative years I would particularly like to single out Surrey Jacobs and Helen Aston for their advice and information on aquatic plants over a period of many years.

The lists of experts and enthusiasts who have helped me with information on wetland animals and their ecology must run into thousands by now, and it would be a daunting task to produce an exhaustive list of their names after many decades, so all I can do is thank everyone who has supplied papers, locations, advice, specimens and even sometimes offered their hospitality over the decades. However, I would specifically like to thank Tarmo Raadik, Peter Unmack, Glenn Briggs, Helen Larson and Mike Hammer for keeping me current with recent developments in the wonderful world of freshwater fishes, Andrew Shaw, Gary Moores and Doug McColl for providing me with some special fishes for photography, and Dave Wilson for his generous services as a guide and all-around expert on Top End plants and animals.

For recent advice and commentary on salt lake crustaceans and also some other groups my particular thanks go to Brian Timms, and to Stephen Charas for providing fine specimens of spanner-claw and Murray crayfishes. Thanks also to the various experts at museums and herbariums around Australia, many of them unfortunately anonymous, who have helped out with difficult identifications over the past two years of photography. A special bow must be made to Ted Hamilton at CSIRO Publishing who first suggested that I should write this account of wetland animals from an ecological perspective; Ted has also been generous with advice on digital photography and the best use of associated software over the course of the past year.

Last but far from least, thanks yet again to my wife Jan Ratcliff who has already lived through the writing, photography, editorial and publication stages of a dozen earlier books. Despite the valuable lessons that should have been learned from these her first comment on the concept of this book, which would require weeks of travel on a shoestring budget, was something like 'You've always dreamed of writing a book like this, and you've already done most of the research and photography needed. Go for it.' A year later as I begin to clear up the debris of improvised photo studios and file away avalanches of books and papers, Jan still thinks it was a good thing to have done!

Introduction

In a sense this book has taken more than 50 years to write, from my earliest formative years collecting, studying and often breeding all manner of creatures that live in water, to scaling up my various interests when I set up Australia's first specialist indigenous wetland nursery in the late 1980s. Originally intended as a way for me to experiment with the many roles and values of plants as animal habitat, the nursery also became a useful source of income that helped to subsidise travel and photography in waters further afield.

In turn, writing about some of my earlier work established me as an author and photographer, though the varied books on wetland plants I have had published cause confusion for the many people who don't realise that my primary background and qualifications are in zoology, rather than botany. This compact volume is a return to my roots, a book I would have loved to have written 30 years ago though it has benefited greatly from the vast new realms of information that have become available in recent years, and at the hands-on level digital cameras and some fine software can now salvage worthwhile images from all but the most hopeless shots taken a long time ago.

This isn't intended to be a book about particular animal species, but aims to show how ecological patterns, processes and animal interactions merge into a seamless whole, with many common themes that are shared across the entire span of this island continent. Although it is primarily intended as a plain language guide to the animals and their ecology in inland waters, some of the most recent information included should also be useful for specialists looking for an overview of what other researchers are working on in unrelated fields. To keep the text flowing smoothly, scientific names (along with any pertinent comments on these) are grouped in a separate appendix rather than scattered through the text.

Many of the chapters have been written as related essays that can be dipped into in any order, so that readers who are already reasonably familiar with the major animal groups found in wetlands can skip to any one of the later chapters according to their mood or interest. If you don't already have this kind of background start with the six chapters in the first section, as these cover the biology and ecology of the major groups from the hordes of miniscule species that feed entire ecological systems, to the diverse carnivores that help to shape wetland ecology and even the behaviour of other species by their presence or absence.

The second section of the book looks at the nature of the wetlands themselves, and the many ways in which changes (both seasonal and also more dramatic changes over longer periods of time) drive many shifts in aquatic animal communities, and in turn their interactions over longer periods of time. The first two chapters in this section focus on changes in the physical environment and the ways in which wetland animals adapt to these, including the reasons that some animals are limited to particular types of wetland. The next two chapters focus on the interactions between the animals themselves, from the changing species and relationships in a shallow lake as it refills after drought to the ways in which predators affect their prey.

The third and final section looks at particular groupings of wetlands to emphasise other types of physical change and divergence, with the first of these chapters following the overall path of water from the highest mountains to the sea and even underground, showing how this path affects the qualities of the water itself as it dissolves various chemicals on its way downstream. The following chapters group wetlands in more traditional ways, from faster-moving streams to sluggish inland rivers, marshes and swamps, lakes both fresh and salt, the melting pot where many inland waters blend into the sea, and finally the problems and also promise associated with wetlands and dams created by humans for their own, often inscrutable purposes.

Part 1
The diversity of wetland animals

In a broad sense ecology is the study of food: the things various animals feed upon, the means and specialisations by which they feed, and their diverse interactions in the pursuit of food. For the great majority of living things on this planet, food begins with plants and other photosynthetic organisms, which use solar energy to convert simple chemicals such as carbon dioxide and water into more complex, energy-storing materials. In turn, these are fed upon by a diverse array of tiny invertebrates, which become the prey of larger animals.

The first six chapters of this book look at the great range of animals that live in wetlands, starting with the invertebrates. In a sense, 'invertebrates' is a grab-bag term that covers the greatest diversity of animal life on this planet, simplistically contrasting them with vertebrates: animals with backbones, including ourselves, fishes and frogs, snakes and skinks, platypuses and birds. Despite their relatively impressive size there are only a few tens of thousands of species of vertebrates, in contrast to the many millions of invertebrates. And as many of the major invertebrate groups make their appearance in this book it will become obvious that their basic body plans are much more diverse than those of the vertebrates.

Many invertebrates are too small to need any kind of solid body structure, but the most successful wetland groups both in terms of abundance and species numbers are all arthropods. Like us, all of these animals are more or less symmetrical, with one side of the body the mirror image of the other, and eyes and mouthparts at the front end of the body. Their hardened skin is known technically as an exoskeleton, firm enough for their internal muscles to attach to, but is not particularly flexible, so their limbs and mouthparts must be hinged and jointed or they wouldn't be able to move. These joints can be intricately arranged, for example the hundreds of finely divided rings that allow the long, whip-like antennae of a crayfish to flex. Some arthropods have rigid exoskeletons, but in others it may be so soft and thin that it isn't always obvious it is even there.

In wetlands the most abundant, diverse and ecologically significant arthropods include the insects, which are looked at in the second chapter, and their more ancient relatives the crustaceans, which are so widespread, sometimes unbelievably abundant, and yet so generally unfamiliar to most people that they merit a chapter of their own. Other arthropod groups, including spiders, centipedes, mites and scorpions, are less common in wetlands, although the ecological roles of some of these are touched on in later chapters.

1

Invertebrates: the crustaceans

Crustaceans range from tiny and primitive aquatic forms which have changed little over hundreds of millions of years, mainly because they are so well adapted to diverse types of aquatic environments, to highly specialised creatures such as crayfishes, which include by far the largest freshwater invertebrates in the world. The original crustacean body design is a series of fairly similar segments trailing away from a more or less distinct head, and many of these segments have gills attached.

Gills are finely divided structures found in many unrelated types of animal, and are basically structures where the blood runs through the fine blood vessels known as capillaries, coming in close contact with water. Blood that is low in oxygen absorbs this essential gas by diffusion, and as it also usually carries considerable amounts of carbon dioxide (a waste product from the use of oxygen) this is released into the surrounding water at the same time. The re-oxygenated blood is carried back into the body, and distributed as needed. Gills may also act as feeding rakes used to capture tiny prey.

Many other crustacean groups have developed more complex body plans, with different segments fusing or becoming specialised in other ways to create animals in which the original segmented design can be hard to see. The external shells of water fleas and seed shrimps surround a body which may still bear some traces of the more ancient body plan, while in the larger crustaceans such as shrimps, crabs and crayfishes the head and front part of the body has become a single structure with the gills and the tiny crustacean brain enclosed, with separate external limbs for walking and feeding.

In general, the closer any crustacean comes to the ancestral form, the more segments with similar limbs-cum-gills it will have. Fairy shrimps (Anostraca: see

Unnamed species of fairy shrimp from an inland drain.

also chapter 18) are good examples of this basic but still very successful design, and these include some of the larger inland crustaceans, although most of them are small. Fairy shrimps are mostly found in ephemeral waters which dry out at least every few years, with a few specialised relatives that may be extremely abundant in more saline waters. Their eggs are often spectacularly drought tolerant, lasting decades and perhaps even longer in some species, and are so small they may be carried from one wetland to another by the wind.

As crustaceans evolved to fill a variety of new niches, their body plans became increasingly complex, and with 500 million years of experimentation it is not surprising that the more successful groups can't be easily arranged into anything like a tidy pattern or sequence. The classification of crustaceans is among the most complex in the animal kingdom, though to put this into perspective it does sometimes seem that taxonomists who study them are more interested in the obscure and complex, rather than more commonplace aspects of their biology.

The major crustacean groups which follow aren't listed in any particular order, simply because there isn't one; instead, the emphasis here is on ecological importance – and some of these animals are lynchpins without which most freshwater and inland ecosystems would simply collapse.

Copepods

Of all the diverse invertebrates found in fresh or saline waters, copepods are probably the single most important animal food source in most marine and also many fresh waters. Even though these may be the most abundant group of animals

Copepods from a shallow, ephemeral pool.

on Earth, with an estimated 1 370 000 000 000 000 000 000 floating around in the open waters of the world's seas at any one time, most of them are so small and unfamiliar that the entire group doesn't have a common English name!

The copepods found in fresh or inland waters aren't particularly diverse, though most large water bodies will usually have several species present, and each of these will have its own niche, filtering different sizes and types of microscopic food from the water. Lake copepods are most abundant after drought, hatching in huge numbers from eggs within the bottom sediments as the waters rise, but there are also many other types which are found only in small, ephemeral pools that go through a cycle of drought almost every year.

To keep their numbers up these minute animals must breed prodigiously, living only a few weeks while producing enough young to feed a multitude of other animals, as well as keep their own species going. It isn't an appealing job description, but they seem to get by. Other types of copepod have become parasites, lodging themselves in the gills, skin and even the internal organs of larger animals. The body of such specialised species as anchor worms and gill maggots may be so simplified that the females are essentially a living dart that is plunged into the sides or gills of fishes, so that they are recognisable as copepods only from their trailing egg masses.

Of the three most common free-living groups in Australian inland waters, cyclopoid copepods with their long antennae and pairs of egg sacs are generally found closer to shore. Calanoids with their longer, more segmented bodies and a single egg mass are often creatures of more open waters, while harpacticoids have stubby antennae on an elongated and multi-segmented body, and are more often found on and around vegetation and bottom sediments than free-swimming.

Plants: phytoplankton and zooplankton

Plankton are free-swimming or drifting organisms which can be more or less separated into two natural groups: **phytoplankton** which produce their own food from solar energy by photosynthesis, and **zooplankton** which are the diverse animals that feed on phytoplankton, on other organisms such as bacteria, and in some cases even on each other. The smaller planktonic crustaceans are a vital part of the food chain in both freshwater and marine ecosystems because these specialists feed directly upon phytoplankton, and are fed upon in turn by many larger animals.

Most phytoplankton are much tinier than the colony of the Volvox algae at left (the green sphere with several smaller daughter colonies within), yet it in turn is dwarfed by copepods just 3 millimetres long.

Phytoplankton are mostly single-celled, free-floating, photosynthetic, single-celled organisms. Some of the most abundant groups include the strangely horned and sculpted dinoflagellates, diverse diatoms that look like miniscule jewel boxes with an endless variety of strange shapes and patterns, and many other algal species right up to the relatively gigantic colonies of *Volvox*, a hollow, nearly transparent ball that splits to release many smaller daughter colonies when it is mature.

Not all of these are plants in any sense that most of us are familiar with, as many of the single-celled organisms which swarm in warm, sunlit waters don't fit tidily into any of the older traditional classifications. Slipper animalcules are the classic example: elongate green cells that act like any other form of phytoplankton in sunlight, fixing solar energy directly through their green, photosynthesising parts. Yet if kept in the dark they lose their green colour, and start absorbing more complex nutrients instead, like most single-celled planktonic animals.

Plant or animal, predator or prey, planktonic organisms don't all just drift. Most phytoplankton are such simple organisms that they function well only within a limited range of temperatures and light intensities, and to avoid variations that don't suit them they will rise and sink over the course of each hour, day and season. And zooplankton must pursue their phytoplankton prey, moving up and down in the water column to wherever their food source is most concentrated. As many copepods can move at up to a metre per minute (equivalent to a kilometre every 2 minutes on the human scale), the daily migrations of many types of plankton could be more accurately described as migrations, rather than just random drifting.

Other smaller crustaceans

Water fleas (Cladocera) are more rounded or disc-shaped than copepods, and carry their eggs within a (usually) transparent shell, from which their swimming arms protrude. These look remarkably like a pair of antennae because that is exactly what they are, and later in this book we will encounter many other examples of arthropods that have a much more flexible limb plan than the vertebrates we are most familiar with. Their resemblance to a flea is limited to their erratic swimming action, sinking slightly, then jerking upwards with each stroke of the antennae.

Often found in the same waters as various copepod species, water fleas tend to be much more abundant in smaller water bodies such as ponds and seasonally dry ditches in Australia, though in the Northern Hemisphere they may also be the dominant animals in lakes. Their single eye is not much good for seeing detail, but measures the intensity of light so that they know when to move towards the surface, and when to sink towards the bottom. There are various theories as to why diverse planktonic arthropods (including these crustaceans) migrate between the surface and the mud below each day, sometimes over considerable distances, and though no one seems to have a definitive answer it is believed they may be following the movements of the tiny algae they feed upon, or avoiding the larger animals that hunt them in turn.

Lentil water fleas.

Even though water fleas and copepods may be found in the same body of water, their numbers and relationships often vary dramatically from one season to another, or sometimes even just over a few weeks for reasons no one understands. I have often caught virtually pure cultures of a single species of water flea or a copepod in a pond within a month or two of it flooding, only to find that species virtually gone within a few more weeks, and replaced by some altogether different animal. In other seasons, the same pond may provide different mixes and combinations, and occasionally still other species, though the dormant eggs of all these animals must lie mixed together in the dry sediments between the wet seasons.

Resting eggs don't just survive drought; they can also pass through the digestive system of birds without harm, which is why many of these animals can be found in almost any water body where birds have visited. They are also very long-lived, and some species have been hatched from dry mud collected from ponds which have not filled for centuries. Like copepods, water fleas feed on a range of single-celled algae and bacteria, and some will also browse on vegetable detritus, presumably for its bacterial content, as it seems unlikely that such tiny animals would be able to digest plant matter themselves.

Seed shrimps (Ostracoda) look something like a seed, and something like a tiny swimming clam. Some species thrive in shallow pools which dry out within several months (or even just weeks), because their two half-shells can close together

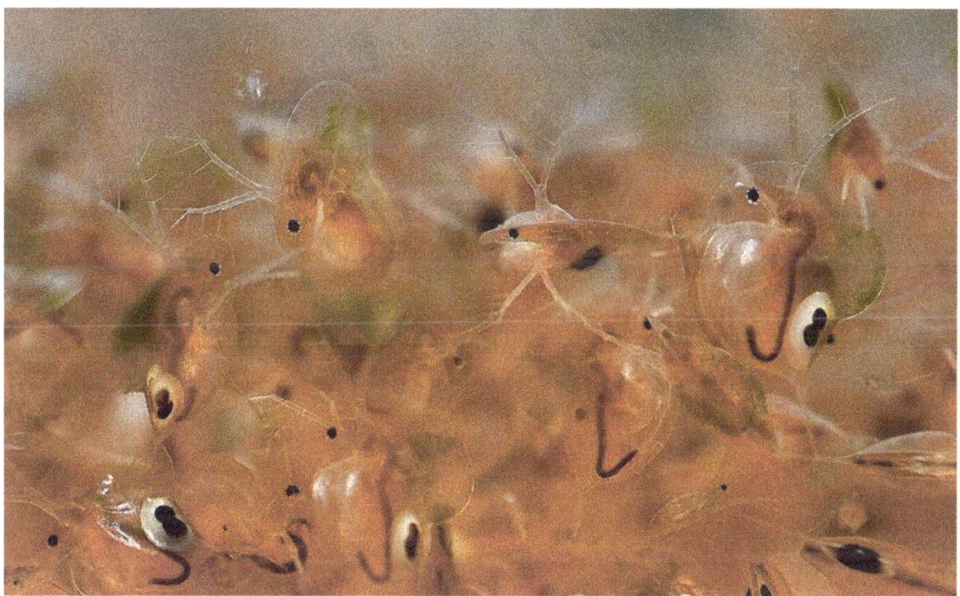

Helmeted water flea females, many with resting eggs forming within.

tightly enough to protect the living animal from drying out. This gives these drought-tolerant animals a head start over other small crustaceans that hatch from eggs, needing at least a few weeks to reach breeding age. By contrast, a dormant population of drought-tolerant, adult ostracods can potentially start breeding within days of a temporary pool filling up.

Other species live in more permanent waters including salt lakes: such giants as the multi-coloured pearl-ostracods, all of 5 millimetres long, and their fresher water relatives the mussel-ostracods, named for their unusual teardrop shape. Most smaller species are less easily identified, though the shells of others are distinctively sculpted or pitted, and these patterns can be used for identification in combination with other details such as whether they have a primitive eye. Some of the more distinctively sculpted shells of some species may be so abundant in freshwater sediments that they can be used as a useful guide to the climates and wetlands of the past.

Clam shrimps (Conchostraca) are generally much larger than seed shrimps, reaching more than a centimetre in length in some cases, but they are also much less often found even in the rainwater pools where the two groups are most likely to occur together. Clam shrimp shells are often sculpted with curving growth lines that make them even more clam-like in appearance, and suggest a similar burrowing lifestyle, but the animal within is distinctly segmented with many pairs of swimming and filtering legs. Like seed shrimps they feed on a variety of algae and possibly also detritus.

Seed shrimps from a shallow pool in a lawn.

Shield shrimps (Notostraca, also known as tadpole shrimps) are among the largest freshwater crustaceans and can reach the length of your little finger, yet these short-lived animals are found only in ephemeral waters along with copepods, water fleas and sometimes fairy shrimps. Their long-lasting eggs may hatch in huge numbers after drought, and as the pools they live in dry out their stranded bodies may attract waterbirds in numbers. With the front half of the body protected by a soft shield, shield shrimps swim rapidly, if rather erratically, with the many feathery pairs of legs which also gather the tiny organisms they feed on. These versatile limbs also function as gills, and even as digging tools when they scour pits into soft sediments in search of food.

Scuds and sideswimmers (Amphipoda) are strongly segmented animals with flattened sides and distinctive hunched backs. If they rear up their overlapping body plates they look like Samurai armour, and with their tiny, glinting, multi-faceted eyes they would be the stuff of nightmare if they weren't so tiny, cute and toothless. The common names refer to their curious, somewhat uncontrolled swimming style as they veer and rock from one patch of cover to another. Most are quite small though some may be as long as the first joint of a finger, and they combine the ability to toddle along on their seven pairs of legs with a sudden,

Shield shrimps.

erratic and distinctly wonky swimming style driven by three pairs of swimming legs that are set too far back for accurate steering.

Amphipods are never as dramatically abundant as copepods or water fleas, but wherever there is plant debris and abundant organic matter they are likely to be present in good numbers; they are also found in a wide range of both permanent and ephemeral waters. Some species have left the water to become the familiar leafhoppers of gardens and sandhoppers that swarm under seaweed at the high tide mark, but they generally prefer wetter places where they survive partial drought by burrowing into organic matter. Others hide among decaying plants or in crayfish

Common sideswimmer.

burrows. Although relatively hard-shelled and evasive, amphipods are most abundant where fishes are absent, or in murky waters where predators have trouble finding them.

Crayfishes, shrimps and crabs

Most of the larger crustaceans are decapods, which means 'ten-legged', though their front legs are often a pair of claws and other legs further back on the body may also be very claw-like. All decapods have a well-defined head and thorax fused together, and females usually carry their eggs under the segmented tail until they hatch. In marine crayfishes, shrimps and crabs these eggs usually hatch into tiny planktonic creatures which go through several distinctive stages until they are ready to settle down to their adult duties. These are often so different from their parents in appearance that they have been given separate biological names for each stage.

Diverse planktonic larval stages don't work for freshwater forms for two reasons, the most obvious one being that the young could easily be swept out to sea in coastal streams, or possibly end up settling hundreds of kilometres downstream from any suitable habitat. The other issue is food: running waters don't have the diversity of minute planktonic organisms that marine decapod larvae feed upon, and nor do still waters such as slow-moving streams and lakes where the water can

Redclaw crayfish.

be so nutrient-poor that light can't penetrate very far (see chapter 16). All of these animals also share a somewhat casual attitude to their limbs, shedding claws or legs if attacked, but able to regrow them gradually with each new moult.

Australian crayfishes are extremely diverse, including by far the largest and the smallest freshwater species in the world and include many of the largest freshwater invertebrates anywhere, in a league of their very own. In ecological terms the major species groups have such well-defined habitat preferences that their presence can tell us much about the nature of a wetland, from the likelihood and regularity of drought to water quality. In fact they could almost be used to define some types of wetland, and will appear in various later chapters as indicators and guides to the nature of the worlds they inhabit.

Unlike most other larger invertebrates many crayfishes can travel overland during wet weather, usually at night as they escape drying pools in periods of drought, and the smooth crayfishes seem to take a particular pleasure in doing so. Apart from this being an exasperating habit for anyone trying to farm them, it has also allowed several species to escape and naturalise in places where they can potentially disrupt entire aquatic ecosystems. The bulk of their diet is usually fresh or decaying vegetable matter, mixed with the minute life forms which break it down, and crayfishes will also scavenge dead animals of any size.

Smooth crayfishes include the most familiar freshwater species: the common yabby of the south-east, marron and redclaw. Some people use the term 'yabby' for all types of crayfish, but this is a south-eastern name that should be reserved for the smooth crayfishes of this area. All smooth crayfishes are fast growing compared to most other crayfish groups, and with their relative tolerance for heat and poor water conditions these are the main aquaculture species.

Spiny crayfishes favour cooler conditions and reasonably clear waters, and those that live at more northerly latitudes are mostly found in high-altitude streams. Each neighbouring set of river systems in eastern and south-eastern Australia seems to have its suite of own distinctive spiny crayfishes, often spectacularly patterned with reds, blues and vivid greens, while further south the closely related Tasmanian giant crayfish and its cousins were formerly the dominant stream invertebrates on that island.

Marsh crayfishes with their deep bodies are often confused with land crayfishes, but their tails are not particularly reduced so they can still swim backwards strongly, an indication of their more aquatic life cycle. Although usually found in shallow, ephemeral wetlands that may become quite saline as they dry out, marsh crayfishes burrow readily in adverse conditions, but feed and breed during the wetter months while their ephemeral wetlands and streams are still full.

Land crayfishes are the most specialised and peculiar species, rarely seen as they don't leave their burrows by day. These are primarily scavengers and foragers with distinctly shrunken tails, which combined with their exaggerated claw sizes gives them a very strange appearance if you are familiar only with the more aquatic species. Their mud chimneys may be a prominent feature around springs and in open ferny gullies, with the burrows dropping down into the water table below.

There is no great difference between shrimps and prawns, but in Australia the latter term is usually used for any that are large enough to be worth eating. Diverse species of freshwater shrimps are found across the mainland, but the glass shrimp of eastern Australia and Tasmania is the most widespread and abundant. An adaptable animal, it is found in slower-moving streams, lakes and backwaters, and even in ephemeral streams that may dry out to just a few waterholes. Like most other larger crustaceans glass shrimps are very sensitive to a range of toxins, so their presence is usually a sign of good water quality.

Freshwater prawns (also known as long-armed prawns) are not all that different in general appearance to shrimps though they are often much larger, and the territorial males have long, slender claws which can be longer than their bodies. The tropical cherabin is also found in South-East Asia, but the southern freshwater prawn is comparably widespread in its various forms, extending through the Murray–Darling well south into Victoria. These are animals of slow-moving waters, usually found near cover including dense vegetation and masses of snags.

Compared to many other warmer parts of the world, there are almost no inland crabs in this country, though the brownback or freshwater crab is found across much of the top half of the continent in a remarkably wide range of habitats from permanent coastal streams to saline inland watercourses that may remain dry

Cherabin.

for years. Although it is currently regarded as a single species, the great diversity of habitats it is found in suggests considerable genetic variation, and it is likely that future studies will turn up many cryptic variations including new species, as is already happening with some of our smaller freshwater fishes. The estuarine crab is a tiny species of false spider-crab, usually found in estuaries or not far from the coast, but as it also enters much fresher waters it may sometimes also be found a long way inland in slightly saline lakes and streams.

Moulting and growth

The ghostly, shed skin of a damselfly nymph with the newly moulted animal behind.

The arthropod exoskeleton isn't very elastic, so it must be split and shed (moulted) whenever the occupant grows too large for it, by which stage it presumably feels something like a pair of very tight pants. Moulting is a complex process involving splitting the old skin along naturally weak seams, with the soft and vulnerable animal within wriggling its way out where these separate. Once the old shell is shucked off, the shell of the newly emerged animal is still relatively soft and must be inflated to a larger size while it remains flexible, allowing space for the animal to grow until the next moult is needed. For the few hours (sometimes days) it takes for the newly inflated shell to harden, the animal is more vulnerable to predators than usual and will usually hide for this time.

It is also an opportunity for larger and harder shelled species, including some crayfishes and crabs, to mate. A male brownback crab may court the female for some time before moulting, mating with her a few hours after the moult, and then guarding her until her shell has hardened. In most freshwater crustaceans the animal that emerges from its old skin doesn't change much except in size, and this is also true for many insects with a juvenile aquatic stage, but other insects go through a much more dramatic moult between juvenile and adult, as will be seen in the next chapter.

A freshwater crab slips backwards out of her old shell; the new and elastic shell is literally blood-red as she works to inflate it before it hardens.

2

Invertebrates: the insects

Most of the thousands of species of crustaceans are aquatic, and they are found from inland waters to the depths of the oceans. By contrast, nearly all insects live at least part of their lives on land or in the air, though many groups have a larval stage that lives in fresh water, and others have moved back to the waters as adults though they can still up-wings and leave for new waters at any time. Wetland insects are often among the most abundant invertebrates in many inland waters, yet the number of species in Australia can be measured in the thousands, compared to the hundreds of thousands on land.

You are likely to find some aquatic insects or their young anywhere that water accumulates or flows through, from ephemeral pools to streams and pools, hidden in silt and organic matter, or swimming free in lakes – but always in the relative shallows. Not surprisingly, insects are probably a significant food source for many other wetland animals, and emerging flights of some of the more prolific species such as mayflies may attract predators from great distances, including terrestrial birds.

The insect body plan

As with the crustaceans, the basic body plan of insects was originally made up of similar segments, and traces of these can still be found if you look closely, especially in their larvae. The insect head is well defined, followed by a mid-section known as the thorax with six jointed legs (a pair on each of the three segments here), and a tail or abdomen which is more obviously segmented than the other two sections. As in the case of many other bilaterally symmetrical animals the

Segmented predatory larvae of a dytiscid waterbeetle.

head end handles feeding, sight and coordination, and in most insects the tip of the tail is where the reproductive organs are located.

The most obvious advantage insects have over crustaceans in moving from one wetland to another is flight, and many aquatic species are excellent colonisers of newly formed waters. Most flying insects have four wings that are also attached to the thorax, and come in two basic models: those that can be folded away when not needed, and those that remain fixed in place at all times. Wings that can be folded away and covered by protective wing cases are most useful for longer-lived adult insects, and these cases are particularly noticeable in beetles as they are often brightly coloured or patterned.

Only around one in a thousand living insect species has wings that can't be folded away; this suggests that protecting your adult wings is a good idea in terms of long-term survival, though many of the species that haven't bothered evolving ways of protecting their flying wings are wetland species such as mayflies that will need them for only a few days. In longer-lived groups with rigid wings such as dragonflies, the combative males often show considerable wing damage by the end of their life, though their wing structure is so strongly braced they can still fly well even with surprisingly large pieces missing.

It is possible that many of today's aquatic insects never completely abandoned the shallow waters and marshy soils where they first evolved, hundreds of millions of years ago. This would explain why their larvae often have gills, but not why all adult insects breathe through a network of tracheae (pronounced 'trak-ee-yay') that form an intricate network of fine pipes and tubes through their bodies, allowing air to circulate internally. Tracheae also explain why most insects don't have red blood: the pigments in haemoglobin and related compounds are there to attach and carry oxygen, whereas in insects that can breathe air this is delivered directly from the outside world, into all parts of the body.

Terrestrial insects drown under water if the openings (spiracles) into the tracheal network are sealed by water. Tiny aquatic insect larvae such as phantom midge larvae don't need spiracles because their bodies and skin are so thin that oxygen and carbon dioxide pass in and out fast enough, but as a result they tire out quickly if forced to wriggle too often, relying on air-sacs at either end to support them passively in the water instead. Larger types of insect larvae are too large and thick-skinned to get away with this, so they may have 'tracheal gills', which in damselfly larvae are just three sheets of tissue attached to the tail; these are so thin that gases can move through their surface. To improve the rate of flow, the damselfly larva pulses its body regularly, creating a flow of water past the tracheal gills; these can be seen clearly in the moulting photo on page 15.

Many other terrestrial groups of insects have moved back into the waters, and as they have no well-evolved and time-proven methods of breathing underwater, they just carry a bubble of air with them. Some trap their bubble under the shields that protect their tucked-away wings, and have to return to the surface to recharge

A so-called glassworm, larva of a phantom midge supported by silvery air-sacs at either end.

A backswimmer bug with prominent compound eyes, a silvery physical gill further back.

their supply every few minutes. Others go a step further with a 'physical gill', a bubble held in place by hairs so that much of it is in direct contact with the water.

As with true gills, this simple bubble arrangement allows oxygen to percolate in from the water, while carbon dioxide leaches out. If the insect is small enough and the water is cool enough to hold a generous supply of oxygen, it can stay down for hours, but the bubble will gradually shrink to the point where gas exchange is no longer effective, and the insect will have to return to the surface every now and then to top it up. For larger insects such as diving beetles the air bubble is mostly trapped under an impervious wing, so diffusion would never provide enough oxygen. These insects are more closely tied to the shallows, and must use more of their energy in commuting to top up their air supply.

The compound insect eye is another development that makes insects much more effective hunters than most crustaceans. Essentially this is a series of honeycomb-like lenses, often forming beautiful patterns, and focusing light in to a central point as in the single-lensed eyes of many larger animals. I remember unimaginative interpretations of what the insect eye can see in some ancient and rather boring books on natural history from when I was young. Not surprisingly, their idea was that an insect with a hundred eyes must see a hundred separate pictures, but images coming through the eye are interpreted in the brain and are run together into a useful picture, regardless of how distorted the original information may be.

Thus, newly hatched chickens fitted with glasses that turn everything upside-down adjust their mental image so they can live a perfectly normal life – until you remove the glasses and they have to learn to see things the right way up again. In the same way each separate lens in the compound insect eye adds another piece of information to the picture, yet in the brain it must all become a seamless whole. And if you consider the wrap-around lenses of dragonflies which give them something close to 360° vision, these spectacular fliers, aerial acrobats and hunters may well see the necessary parts and colours of their world far better than we do.

Wetland insects: the major groups

There are many small and often obscure insect groups that are found in and around wetlands at some stages of their lives, but however interesting these may be they are just incidental flickers in the overall ecological panorama. These include the larvae of some moths, lacewings, spongeflies and craneflies, not to mention many small and inconspicuous beetles that bumble and crawl their way through streamside vegetation and shallow waters. For this reason, the brief summary which follows concentrates primarily on the major groups of wetland insects, those which overwhelmingly dominate in many different aquatic environments, both in water and above it.

Of these, dragonflies and damselflies (Odonata) are the most familiar and conspicuous wetland insects. Dragonflies are mostly recognised by the way their wings remain outstretched when they settle, while in most damselflies the wings are usually folded together above the body once they have landed. These are theoretically among the most primitive insects, in the sense that they haven't bothered improving their design over the past 300 million years, but then they haven't had much need to.

Black-headed skimmer.

Young dragonflies and damselflies are stealthy underwater carnivores (see chapter 10), and like the larvae of many other insects which spend their prolonged youth underwater are known as nymphs. In plant-rich wetlands these are among the most effective predators on many other aquatic invertebrates, tadpoles and smaller fishes, combining slow cunning with a huge appetite, and with the potential to reshape the ecological balance of entire underwater communities. And these superbly designed carnivores can be delivered into any suitable water body, however isolated, by their flying parents.

In Australia dragonfly nymphs are also known as mudeyes, but damselfly nymphs don't have a specific name because they are rarely noticed, even though they may be among the most common underwater insects. If you look closely into weedy tangles in a clear pond, you may notice the regular jerks of thousands of these tiny, goggle-eyed monsters as they keep up the flow of water past the three-vaned gills fanned out at the tip of their tails.

Gills of the generally larger mudeyes are inside their rear end, within what is politely known as their rectum, and the water they inhale through this can be blasted out to propel them abruptly forward in a form of rocket propulsion. Some species will also use these bursts of water to chase prey, as I once found when I added several backswimmers to an aquarium with a very hungry mudeye. Its bursts of speed were impressive, and within less than 30 seconds it had cornered and seized a backswimmer after perhaps 10 or 12 rocket bursts.

Although the wings of dragonflies can't be tucked away or even folded, they can still outfly almost everything else in the skies, hovering, reversing and somersaulting at a speed unsurpassed by any other small animal. All of this speed and manoeuvrability comes with drop-dead good looks, plus an abundance of attitude. Although it is still regarded as unscientific to attribute any feelings or other human qualities to other life forms, I have no doubt that dragonflies enjoy showing off in their own, primitive fashion.

In turn, these diverse and attractive attributes seem to have created a unique mindset among the scientists who create their common names, of a more elevated and poetical order than most. Flicking through Theischinger and Hawking's definitive book on Australian species, and choosing just a few that particularly appeal I find whitewater rockmaster, bronze needle, green-blue threadtail, powdered wiretail, waterfall redspot and jewel flutterer. Such names aren't just an Australian phenomenon; for example the Dragonfly Society of the Americas also puts a lot of thought into such fine and descriptive names as gilded river cruiser, black-mantled glider, beaverpond basket-tail and cardinal meadowhawk.

From the Philippines comes the radiant gossamerwing (a concise but accurate description), the golden flatwing from Belize, the twister from Madagascar and orange emperor from South Africa. And although names such as the hairy hawker of Ireland and the common winter damselfly of Crete may seem a let-down among such exalted company, even these have a certain cachet when considered objectively. My take on this matter is that, with such distinctive and descriptive common names, we could dispense with scientific names for damselflies and dragonflies entirely.

The purpose of all these bright colours and fine aerial displays is reproduction, and there can be no doubt that these vigorous flyers and active aerial hunters can see very well, with their immense, multi-faceted eyes that make up most of their heads. Males spend much of their time driving off rivals and, once a female has been attracted and consents to mating, will invest much time and effort in keeping her. Although the male's true penis is at the far end of the abdomen as in most other insects, before mating he transfers sperm from this to a secondary penis at the base of his abdomen, where it joins the thorax.

The mating pairs fly linked into a wheel, with the tip of the female's tail joined to the secondary penis, and the tip of the male's tail holding onto her head. Some species may fly linked together as the eggs are laid, so the male can pull her back to the surface if she goes too far under while depositing eggs in submerged plants. However, in many other species she will fly alone, scattering eggs from the air or landing cautiously and just stretching her tail under the water.

Emergence of a dragonfly

The transition from juvenile to adult insect involves escaping from the restricting external 'skeleton' with each upgrade in size. Some insect groups change gradually from one moult to the next, so they don't look all that different from the adult form even in the beginning, and there is usually no great surprise after the final moult when they turn into winged adults.

Dragonflies make a pronounced change of this kind when they leave the water, eloquently described as emergence. Although it has often been assumed that mudeyes usually emerge by day in the cooler climates of southern Australia, in my experience this is a rare event, and they are far more likely to wait for a warm night when they are safe from most potential predators. The images here show the development of a blue-spotted hawker from when it first crawls out of the water in the early evening.

The old skin is split not long after dark, with the emerging dragonfly hanging weakly for a few minutes, then abruptly flipping itself upright and pulling its tail out of the empty mudeye case. Its first priority is to inflate the new, wrinkled wings over the next 2 hours while they are still soft, and once these have reached full size and are hardening, the tail also expands to its full length. The newly emerged adult can't fly until it has warmed up in the early morning sun, shivering intermittently to get its flight muscles working, but then it suddenly lifts into the air and is gone.

There are many families of true flies other than those which are just household pests, and many of them have aquatic larvae which are active filter-feeders, collecting tiny algae, bacteria and other small stuff in intricate nets of fine hairs. Even mosquitoes are flies. It is only the female that needs blood as a high-protein nutrient source for egg production. Her eggs are laid in tiny, floating rafts, preferably in small pools of water as the larger and deeper a wetland is, the more chance there is that it will be home to predatory fishes or insects – all of which love mosquito larvae. This is why spray programs to control mosquitoes in lakes and wetlands near developed areas often have little effect; the mosquitoes are more likely to be breeding in old water tanks and even the odd ice-cream container someone has accidentally dropped under the deck.

If you can overlook the obvious disadvantages of having mosquitoes around (apart from the itching their bites cause, they are potential carriers of many of the world's most serious diseases), their larvae are entertaining to watch under high magnification. As they sink slowly, their mouthparts open to reveal lips that look like floppy mops, instead of the more formidable grinding mandibles of more

Bloodworms.

actively hunting insects. Every minute or two an actively feeding larva will twist to mop its bristling tail, or run the mop along its body to collect whatever microscopic life has made contact. At other times they lash and twist along to a new position, usually triggering off the same crazed writhing in the other wrigglers they pass, then drift again with limpid, vacuous eyes.

Biting midges are related to mosquitoes, but bloodworms (chironomid larvae) are the young stage of non-biting midges which could be mistaken for mosquito wrigglers, if it was not for their distinctive colouration. These are often abundant in oxygen-poor environments, and their bright red or orange blood makes them among the most conspicuous insect larvae, especially when they whip and coil frenetically near the surface of the water. Most of these larvae feed on the algae and bacteria associated with muddy detritus, and as they are also readily eaten by most fishes they are usually found in large numbers only where the water is too murky for predators to find them, or there isn't enough oxygen for most other animals.

Other aquatic fly larvae may be much less visible even in clear waters, particularly those of phantom midges (chaoborids; the larvae are also sometimes called glassworms), which are almost completely transparent. The most conspicuous feature of their bodies is the air-sacs at either end, which shine silver as the light that passes through them bends out of its normal path (refracts). These are among the most intriguing fly larvae, with their brightly coloured antennae modified to seize and stuff tiny animals into their mouths, yet although they may be present in their hundreds drifting near the water's surface in a still pool, they can only be seen easily when the light slants in at a suitable angle.

Metamorphosis

Many insects (including all true flies) make an abrupt and radical transformation from their larval stage to the adult, known as metamorphosis. For example, the elongated, caterpillar-like larvae (wrigglers) of the mosquito are designed to be efficient filter-feeders, spending most of their time near the surface breathing through a short snorkel at their rear end, but diving to the bottom of the pool at the slightest disturbance. When the time comes to change into an adult the larva takes on a very different, more blocky shape called a pupa, and using that as a kind of template it basically dissolves inside and reassembles itself into a very different guise: the flying mosquito, designed primarily to mate and breed.

Waterbeetles also metamorphose, but the larvae of many of these are highly effective carnivores as can be seen from their impressive jaws (see chapter 10). To change into adults these larvae usually have to leave the water, changing into pupae in some secret place, and emerging later as very different-looking carnivores which may return to the pond where they were born, or have the option of flying to new waters.

Filter-feeding mosquito larvae (wrigglers).

Mosquito pupae with the outline of the future legs already visible through the thin husk.

Many other insect groups are also popularly known as flies, though these have been following their own distinct evolutionary lines for hundreds of millions of years. Their superficial resemblances aren't just in the appearance of the adults, as many of these also have larvae that are popularly referred to as maggots, which also metamorphose dramatically when the time comes.

Mayflies (Ephemeroptera) are as ancient a group as the dragonflies. Their larvae are among the most abundant insects in many rocky streams, and sometimes also in lakes where other species burrow in silt. The nymphs of most species feed mainly on algae and plant detritus, though there are also a few minor predators among them, and in turn they are an abundant food for diverse aquatic carnivores. Fishes feed readily on both juveniles and the adults, and carefully tied imitations of mayfly larvae are often used by flyfishers, especially during the so-called hatch when they emerge from water.

At this stage the larvae drift to the surface, and the first of the two winged adult stages emerges: the sub-imago, with leaden-looking wings which are imitated by 'dun' trout flies. This takes shelter in streamside vegetation for a day or so, before moulting again to reveal fully transparent wings, and the final adult stage takes off on its mating flight. Although the adults live only a few hours or sometimes days, they may take to the sky in such huge numbers on these flights that they create an impression of billowing smoke.

The aquatic larvae of stoneflies (Plecoptera) are primarily carnivores found in rocky streams, with two-pronged tails rather than the three prongs seen on most mayfly nymphs. They are also longer lived than mayflies, and don't emerge in the same huge swarms. Their wings are the next evolutionary step from the relatively rigid wings of mayflies and dragonflies, as they can be folded to allow adult stoneflies to take shelter among rocks, or even burrow into soil or plants.

Adult caddisflies (Trichoptera) resemble drab moths with unusually long antennae, but their larvae are among the most diverse and intriguing aquatic insects. Most of them are rarely noticed unless you spend some time sifting through sediments and sunken organic detritus, as they are mostly hidden within a cocoon or case which protects and disguises them as they move around. These cocoons are constructed in many different ways depending on species, the simplest being just hollow sections of reed the animal can retreat into when disturbed, but other more elaborate cases are made of silk and (depending on the species) may be decorated with stones, various types of vegetation, or even formed into the shape of a snail shell.

Some larval caddisflies are passive filter-feeders, living within an elaborate, anchored trap net that faces into any current there may be, and harvesting whatever smaller creatures are caught in this extension of their home. Many others are scavengers, living on decayed organic matter found in the debris they forage among, but there are also many families that are active predators with impressively large jaws. As most caddisfly larvae are sensitive to a range of environmental problems, their presence is usually regarded as a sign of good water conditions, especially in streams.

The elegant, silken cases of a conoesucid caddisfly larva.

Beetles occupy a diverse range of niches in and around water, and most of the larger aquatic species also have aquatic larvae. However, there are also many small and even some tiny species that live on the edge of streams and are rarely seen unless you are deliberately searching for them. Many of these can swim and even dive in a clumsy fashion if disturbed, but they play only a minor part in the overall ecology of most wetlands and are not considered in any detail here.

Beetles (including the larger aquatic species) are surprisingly good fliers considering that their front pair of wings have turned into hardened wing cases that protect the remaining pair of flying wings. And the truly aquatic species are also strong swimmers, with mats of hairs that lock together when they push forward, flattening back against the legs as they are pulled back for the next stroke. This is easy enough to see on the larger diving beetles, while on smaller species such as whirligigs the hairs are so evolved that they look more like paddles as the beetle thrusts forward, folding back to form a smooth sheath as the legs are pulled forward for the next stroke.

The larger and truly aquatic dytiscid waterbeetles are active predators both as adults, and also in their larval stage, which looks a little like a flattened caterpillar but with massive piercing jaws. While the adult beetles actively hunt and seize their prey with their front legs, and chew them up with their mandibles, their larvae are among the most effective and voracious stealth predators, with some unusual specialisations that are discussed in chapter 10.

Great waterbeetles.

Whirligig beetles are a separate family, though their larvae are similar to those of dytiscids in many ways, and the two groups probably share a common ancestry. Adult whirligigs can dive, but they spend most of their time on the water's surface in groups, idling around each other as they wait for smaller animals to fall onto the surface. Their lives are closely tied to the interface between water and air, and their special adaptations are looked at in chapter 15.

The lifestyles of adult aquatic bugs are more diverse than those of aquatic beetles, but like dragonflies all of them go through a gradual series of larval stages that look increasingly like the adults, rather than changing abruptly through metamorphosis. One feature that helps define the group is their specialised sucking mouthparts which allow some terrestrial species that feed on plants to become serious pests of agriculture and gardens, but many aquatic bugs are predators which may inject venom or digestive juices into their prey.

As their mouthparts are too specialised to capture prey, the front legs of many water bugs have developed into seizing limbs that look a lot like the mandibles of other carnivorous aquatic insects. These include the water striders, the most conspicuous aquatic bugs which skate in groups across still ponds and backwaters (see chapter 15), looking very strange for an insect as they seem to have only four legs at first glance, with the front pair being folded up in front of the face. Some inland water striders don't mind a hint of tidal influence in the waters they skim, and with their surface-dwelling habits it is not surprising that more specialised members of this family are among the very few insects found on the open sea.

Creeping water bugs; those impressive 'jaws' are actually highly modified legs.

Among the most widespread groups in relatively still waters are the backswimmers, with oar-like limbs used to maintain an upside-down position below the surface. These are easily recognised by their silvery, cigar-shaped bodies from above, and bright red or pink eyes when seen from closer up. Their larval stages don't look all that different from the adults though they are often a paler colour. These are active carnivores feeding on small invertebrates of all kinds, and should be handled carefully as they have a sharp sting comparable to that of a bee in intensity.

Waterboatmen are superficially similar to backswimmers but swim with their green, camouflaged backs upward, and feed on decaying vegetation as well as any smaller organisms they can catch while browsing through organic matter on the floor of shallow, relatively still waters. While many fishes avoid eating backswimmers presumably because of their bite, waterboatmen are abundant only in waters where fishes are absent, or which are too murky for predators that hunt by sight.

Needle bugs are often among the first colonisers of new farm dams, sometimes arriving as soon as there are enough smaller insects present to feed them. This also makes them an excellent means of transport for some parasitic organisms, particularly water mites (see the next chapter). Despite their slow and clumsy rowing motion when they swim, these and their broader shouldered cousins, the so-called water scorpions, are skilled ambush predators and can even capture small fishes, seizing with the forearms and then stabbing with their pointed beaks.

Giant water bugs are impressive-looking predators which can be as long as your thumb, with the male carrying the next generation as eggs implanted on his back. Creeping water bugs look similar but are much smaller, with an even more

Waterboatmen bugs.

Needle bugs.

impressive pair of front legs that look more like giant fangs, well adapted to holding even quite large prey. This is one of several aquatic bug families in which the males produce curious squeaking noises by rubbing a leg across their head to attract potential mates, though I have also heard waterboatmen making similar sounds in aquaria, when being pursued by a fish.

3

Other invertebrate players

Crustaceans and insects may be the most abundant freshwater invertebrates, but with their shared arthropod heritage of a jointed external 'skeleton' and related features they cover only a fraction of the range of invertebrate diversity. Many other invertebrate groups evolved alongside the arthropods around half a billion years ago, all of them already clearly different from each other in many aspects of their biology, and a few species of many of these primarily marine animal groups have also moved into fresh waters. Some of the more obscure species are mostly interesting for their evolutionary history, but other unfamiliar groups of invertebrates can be as prolific as crustaceans at times.

Miniscule worlds

Many groups of wetland animals are too small to be seen without a good microscope and few of these are included in this book for this reason, yet some are so widespread and sometimes abundant that it would be hard to write anything about the diversity of wetland animals without at least a brief look at them. As they are so unfamiliar and even bizarre-looking when seen under a microscope, I like to refer to them by the old name 'animalcule' used by early microscopists when they still had no idea what they were looking at, and the strangeness of these tiny creatures only became more apparent as more was learned about them.

Of these, the single-celled creatures commonly known as protozoans are often among the most abundant planktonic life forms, though others such as bell animalcules remain attached to some kind of surface by a slender and flexible stem for most of their lives. Although each protozoan is technically just a single cell, its

A colony of bell animalcules on a mudeye.

internal arrangements are often complex with various structures called organelles visible under powerful magnification, each organelle carrying out specific functions just as organs do in larger and more complex animals.

Most protozoans feed on organic particles and on tinier life forms such as bacteria, but others hover somewhere on the borderline between plant and animal. Consider slipper animalcules, fast-moving and elastic creatures driven by rippling hairs, obtaining their nutrients directly from sunlight like any self-respecting plant. Yet if kept in the dark they willingly abandon their photosynthetic lifestyle, and take to feeding directly on particulate matter like many of their relatives.

Wheel animalcules are particularly abundant in shallow, ephemeral waters (see chapter 7), but the contrasts between these and protozoans are so great that, considered together, these unrelated groups make a striking example of the diversity to be found in microscopic animals. Individual wheel animalcules are often no larger than many of the protozoans they live among, yet each is a complex animal made up of *hundreds* of cells which form into various organs just as in all larger animals. This might seem unremarkable except that, unlike most other animals (which can vary considerably in size and cell count), each specific species of wheel animalcule comes in just a single model, with precisely the same number of cells as its siblings.

Molluscs

The molluscs of inland waters (whether fresh or salt) are nowhere near as abundant or diverse in as in the sea, and their shells are often much flimsier and less ornamented than those of marine species. This is partly because the calcium needed to build a really solid shell is not as reliably available in fresh waters, and though there are many other species such as land snails and slugs that find adequate amounts to keep building their shells in even more calcium-depleted environments, the inland species are mostly restricted to reasonably permanent waters.

Freshwater snails are mostly small with fairly plain shells, and the smallest of them have almost no shell at all, so they look like tiny slugs wearing an inadequate sunhat. Snails may be abundant in some slow-moving waters, with their gelatinous

Pea mussel shells washed up on the shores of a shallow lake give some idea of their abundance in the waters further out.

The nature of water: O_2, CO_2 and breathing

Aquatic plants are often assumed to be the main source of dissolved oxygen in fresh waters, and in some exceptional cases such as the crystal clear waters of Ewens and Piccaninnie ponds in south-eastern South Australia it is even true. Here, the constant flow of lime-rich water produces dissolved carbonate ions, which provide reliable supplies of the carbon dioxide plants need for vigorous growth. Combined with intense light this allows plants that would normally be restricted to the shallows to grow several metres deep, and on sunny days countless streams of oxygen bubbles can be seen rising from them.

In other waters most oxygen comes from water movement at the surface, helped along by the slightest breeze, but the amount present is always much less than the 20% reliably found in air. Smaller underwater animals make the most of these relatively low oxygen levels in various ways, or get around the problem by breathing directly from the air above. The smaller an aquatic animal is, the harder it is for it to break through the surface tension layer at the water's surface; on the other hand it also becomes easier for it to get enough oxygen through gills. For larger invertebrates in stagnant waters gills aren't enough, and the common insects in such places will mostly be surface breathers, rippling the surface as they dash up and break through to replenish their supply.

Streams of oxygen bubbles trail up from plants growing in Piccaninnie Ponds.

Less active animals such as aquatic snails can get by with gills as their metabolism is literally sluggish, like their cousins the slugs from which the term comes, though some larger species do take in air directly from above the surface. In oxygen-poor waters animals that live below permanently need haemoglobin-rich blood not that different from our own, as this compound bonds firmly with every oxygen molecule available. True aquatic worms and also bloodworms (aquatic larvae of non-biting midges) are among the most obvious examples, but even water fleas often turn haemoglobin-red if they are crowded into stagnant pools or ditches, where decaying vegetation uses up much of the available oxygen.

Many water snails can breathe air directly; smaller species are more likely to breathe through gills.

egg masses smothering the submerged surfaces of aquatic plants. They are also a significant food source for birds that forage under waterlily leaves, picking various invertebrates off from the underside as they turn the leaves up with their slender toes.

Freshwater mussels are usually only noticed as dead shells in dry creek beds or among the remains of water rat middens, yet there may be thousands buried shallowly in the silty gravel of some shallow sandbars, and they were apparently a favoured food of many Aboriginal peoples despite their bland taste. Though superficially similar in overall appearance mussels fall into two basic ecological groups: those that only thrive in moving water, and those that prefer relatively still conditions and backwaters.

The slower-water species from inland rivers are exceptionally drought tolerant, able to seal their shells tightly and survive for many months if left high and dry, as long as they aren't left exposed to direct sunlight. The behaviour of these sluggish animals is easier to observe than the fast-burrowing tidal bivalves of the sea, with their most prominent feature a muscular, tongue-like white foot which digs into the silt below, then tows the shell downwards after it. At the other end a pair of fringed valves can sometimes be seen, one of which takes water into the animal where anything small and edible is filtered out, while the other returns the filtered water to the outside world.

The larvae of freshwater mussels are parasitic on the gills of fishes, dropping off when they turn into a miniature of the adult. It must be an unreliable way of making your way to new territories, unless you have some inbuilt way of recognising when you are suspended above a suitable patch of habitat. The most abundant inland bivalves are the small, cockle- or clam-like animals usually known as pea mussels, which are mainly found in the fine silt of slightly saline lakes, often forming a densely packed, continuous band many metres wide around the shallower fringes. Although these can be a useful food source for some waterbirds, like many other aquatic molluscs they are also likely to be intermediate hosts for a variety of parasites.

Worms, leeches, sponges and jellyfish

The term 'worm' covers a multitude of elongated, multi-segmented animals, not all of which are closely related. True freshwater worms are closely related to the familiar terrestrial species, many of which are also surprisingly tolerant of prolonged flooding, and include some that are likely to be among the last survivors in polluted waters where oxygen content drops to almost nothing. Their red colouration comes from the oxygen-carrying pigment haemoglobin which also gives our blood its red hue, giving these adaptable animals the ability to breed profusely in environments where potential predators can't live.

Leeches are more distant relatives of worms, most of them specialised for a life hunting diverse smaller invertebrates. The smaller species are flattened and often translucent with somewhat corrugated segments, and are almost invisible in the perennial ponds and marshes where they live, but on a white backdrop they look like some strange kind of elastic jewel. The leeches most people think of first are the blood-sucking species, including striped tiger leeches and horse leeches, able to wait patiently for many months without food, just waiting for a suitable victim to brush past their ambush.

Other peculiar and somewhat worm-like animals include flatworms, amorphous creatures with a flat body and irregular digestive system. These are very simple creatures in many ways, yet they have intrigued biologists for many

Black water worms.

generations with their ability to grow new body parts as required if cut into pieces. As that includes tails growing new heads, their regenerative abilities are far ahead of more advanced animals such as crayfishes, which can only regrow missing limbs.

Other more obscure freshwater groups such as moss animals and the sponges that may cover sticks or the underside of rocks in stream beds are probably derived from marine ancestors, and even the marine sea anemones and jellyfishes have freshwater relatives. These are mostly so small they are rarely noticed, though freshwater jellyfishes intermittently appear in lakes and ponds in huge numbers; this is possibly an introduced species as it seems to favour disturbed and artificial

This hunting leech feeds on tiny animals, not blood.

habitats. Hydras are effectively miniature freshwater sea anemones, slender-bodied creatures that may build up in numbers following a boom in copepod or water flea numbers, trapping their prey with stinging tentacles.

Spiders and mites

Crustaceans and insects may be the most common and diverse freshwater arthropods, but some of their more distant relatives can also be moderately common in wetlands and slower streams in the form of spiders and mites, always conveniently recognised as they have eight legs. Though similar at first glance the semi-aquatic native spiders come from two different families. Aquatic wolf spiders are blunter and heavier bodied, while the more slender and elegant fishing spiders have longer legs, so they can skate across the water surface in pursuit of prey detected by ripples created along the surface.

Both groups will hide underwater if disturbed but don't stay there long, disappearing under floating leaves or running down the stems of deeper rooted plants, and breathing from the silvery air bubble they trap in the fine hairs that cover their bodies. The females carry their substantial egg masses with them as they hunt, and can tackle some surprisingly large prey. The first time I saw a big wolf spider (of a species I had often picked out of nets by hand) seize a struggling fish three times its length, and run up the side before jumping off with it, was an unpleasant shock.

Freshwater mites are mostly small, fat-bodied animals, coloured or patterned in combinations of bright red and orange on black, and this seems likely to be a warning that they are distasteful or possibly even toxic as fishes won't touch them. Unlike ticks and related unpleasant terrestrial creatures, freshwater mites seem to be harmless and bumbling swimmers in still waters and ephemeral pools – unless you are another invertebrate. While the adults feed on planktonic animals, their young are parasitic on aquatic insects, and are carried to new waters by their victims, which include free-flying species such as needle bugs.

Parasites and other strange life forms

Parasites have evolved in many unrelated animal groups, usually by a process of simplification, stripping their bodies and lifestyle down to the minimum needed to drain nutrients from their host so they can later reproduce their kind. As far as is known most types of animals have at least one parasite specific to them, some have several, and as there are also many more generalised parasites there could be more types of these in the world than all non-parasitic animals put together.

The great majority of these largely unstudied and unnamed creatures are invertebrates, with some groups such as protozoans and diverse worms being particularly prone to this lifestyle, and it is likely that many of the things animals do are intended to minimise the transmission and side-effects of their parasites.

Parasitic mites on a needle bug.

Temnocephalans crowded on a spiny crayfish.

Consider a single, all-too-graphic example: damselfly larvae just 2 centimetres long have been found with up to seven roundworms within them, some up to 5 centimetres long, and it would be surprising if these miserable individuals managed to survive long enough to reproduce. Even the massive blooms of new life that appear in wetlands that have refilled after drought, from teeming planktonic animals to the fishes and birds that feed upon them, may partly be an affirmative response to a brief period of relative freedom from parasites.

Reproduction is the most intriguing aspect of the lives of many parasites, because these stripped-back-to-basics animals often have surprisingly complicated ways of passing on their genes, including different life stages that require different hosts. It seems a chancy business, as they will die out if any of their hosts disappear or move on elsewhere, yet there also seems to be an endless queue of additional species willing to take the chance on the parasitic lifestyle.

And in addition to the obvious parasites there are also many other strange creatures variously referred to as commensals, symbionts and by other names that can never be clearly defined or separated, and are not really understood. Some of these are very successful, including the bizarre temnocephalans that may swarm over freshwater crayfishes. Looking like a cross between a squid and a leech, they are reputed to be harmless to the animals that carry them, and it is rare to find a crayfish in most open waters without at least a few of these passengers.

A green fishing spider waits patiently with its front legs resting on the water's surface, ready to detect the slightest ripple from potential prey below.

4
Fishes

For all of the impressive abundance of aquatic invertebrates and their diverse ecological roles, most of these animals are too small or too well hidden to be noticed by anyone who isn't prepared to deliberately hunt for them with fine-meshed nets and microscopes. Casual observers will see nothing but the larger animals, and apart from a few crayfishes and prawns, most of these are vertebrates: animals with backbones, limbs and eyes not all that different from our own.

Fishes are the most familiar and conspicuous aquatic vertebrates, so streamlined and swift that they are a perfect symbol of the freedom of the waters, yet their lives are constrained compared to the ability of many other animals to travel between wetlands. There are many wetlands no fish is likely to ever reach, even in places so close together that frogs regularly cross between them on a wet night, because nearly all fishes need a bridge of water to travel through.

Those in more permanent rivers have fewer problems travelling around, and a few may even climb wet rock faces on their way upstream, but this isn't an option for fishes that have evolved in more sluggish environments such as swamps, heathland pools or high alpine lakes, any more than a rainforest honeyeater is likely to attempt a desert crossing. Such specialisation ties many fishes to the habitats they have evolved in, and as only a handful of natives are known to have some ability to survive drought, few wetlands that dry out even occasionally could ever support fishes of any kind.

The absence of fishes is no bad thing for many other groups of animals, as nearly all native freshwater fishes are basically predators, even including the few that also include some fruit or algae in their diet. The absence of fishes is the reason that ponds that dry out regularly can be among the most productive habitats, and even lakes where fishes have mostly disappeared after a drought may swarm with huge numbers of crustaceans, insects and other smaller life forms as they refill.

My earliest clear memory is of catching young galaxiids at the age of three, using an old paint tin in a shallow stream at the bottom of the garden. More than half a century later I'm still learning about fishes, and it continues to surprise me how little we actually know even about many common freshwater species. This chapter primarily focuses on some of the most ecologically important families and species, but there are also many other groups we know much less about and that may be common enough in some out-of-the-way places: freshwater soles and pipefishes, flagtails, swamp eels, longtoms, garfishes, anchovies and herring are just some examples.

Moving between the sea and fresh waters

The great majority of native freshwater fishes are derived from families that were originally marine. Not surprisingly, some of these still return to estuaries or even the sea to spawn, and many are close relatives of species that live in brackish or offshore waters (tupong, bullrout and various galaxiids along with some gobies and gudgeons are looked at in later chapters). The small size of most such species makes it difficult to follow them through their brief lives, but larger size doesn't

Most inland fishes in Australia are descended from marine families; the gulf saratoga is one of just four primary freshwater species.

necessarily help that much, as the example of the two largest and most widespread freshwater eels demonstrates.

Marbled eels are common along the broad coastal fringe of eastern Australia and can weigh over 20 kilograms, with the largest females so thick-bodied that they look more like flipperless seals than fish. The smaller short-finned eel is even more abundant in a varied range of freshwater habitats through coastal rivers and streams of the south-east. Both species are also common in New Zealand, and there are related species with apparently similar life cycles in northern Australia, Europe and northern America.

These elongated, night-hunting fishes live most of their lives in fresh waters, feeding and fattening over decades until they reach breeding age. Young female eels, unlike the males, are more likely to move inland, well away from estuaries colonising a wide array of wetlands from slow-moving streams to lakes and even farm dams. The smallest eels (known as elvers) can even climb wet rock faces on their inland journey, held on by the surface tension of water alone, so they are among the very few native fishes found above waterfalls. And although they can't swim against strong currents in rapids, they can make their way through fast-moving waters by tunnelling through loose river gravel.

We have a reasonably complete knowledge of their habits and ecology in fresh waters, but the most curious and mysterious phases of the lives of eels take place in the sea. The brief account of their life here has been repeated with increasing degrees of confidence from one book or article to another over many decades, yet the actual evidence for this story is surprisingly sparse and fragmentary.

Once an eel reaches maturity it begins to reduce its digestive system, because it doesn't have time to waste on feeding any more. Instead, these organs add to the fat

Short-finned eel.

which will fuel its long swim downstream, and across the floor of the sea, while maturing the eggs or milt which must be ready at the ultimate destination. No one is certain where they breed, though it is believed to be hundreds of metres deep in the Coral Sea, just as their Northern Hemisphere relatives from both Europe and the USA apparently travel to the Sargasso Sea in the Atlantic. The one thing we are almost certain of is that adult eels don't return to fresh waters, presumably dying after mating, so their long life in fresh waters is in one sense just a preparation for the final mating journey.

In their earliest stages the tiny young eels are wide and flat, and look so different from the adults that they were originally thought to be some very different type of fish and named *Leptocephalus*, now used as a common name applied to eel larvae generally. Gelatinous and transparent, these flimsy creatures drift on ocean currents towards Australia and New Zealand, probably feeding on some of the many other planktonic animals drifting with them. What they eat is a complete mystery as their spiky, forward-projecting, interlocking teeth don't look well designed to capture prey, and the mere handful of specimens that have been captured have no stomach, and nothing identifiable in their gut.

As they near the coast, the larvae shrink and thicken to become glass eels. Hiding in their thousands in estuaries, these still-transparent elvers start to develop darker pigmentation and grow a different set of teeth for a new diet in their future lives, before moving upstream into fresh waters over the course of spring and summer. Moving mostly by night, the schools of young eels may sometimes be so dense that they form a solid black band in narrow places.

These two species are part of a group of 'freshwater' fishes that live several very different lives, the most divergent and specialised of these in the open sea. Despite their large size the life cycle of these once-significant predators of inland waters is as complex as that of many insects and parasites, from a non-feeding breeder on the ocean floor, to a highly specialised planktonic drifter in the open sea, and ultimately as a transparent although recognisable young eel on its way to a new life in a very different environment upstream.

It is also worth mentioning lampreys here, as these are often assumed to be a type of eel due to their elongated shape, and because they also need free access between marine and fresh waters, but these jawless creatures are only distantly related to true fishes. The adults die soon after spawning in pebbly streams, but their young remain hidden among rocks and gravel, feeding by filtering algae and plankton until they are large enough to move out to sea. One poorly known species remains in fresh water all its life, but in their marine stage the other two species are parasitic on sea fishes until they mature, returning to fresh waters only when they are ready to spawn.

Finding your way in murky waters

Water is around 800 times as dense as air, so any fish that swims at any speed must be highly streamlined, which is why so many fishes are so elegantly tapered. It also means that even the slightest disturbances or pressure changes are more pronounced than in air. The lateral line system seen on most fishes is a row of pores running along the side and around the head, combined with embedded sensory receptors with hair-like cilia that move with the slightest changes in pressure, giving a fish a good idea of what is in its neighbourhood even in murky waters. Dramatic changes or jolts near a fish are often a warning of potential attack that the eyes may not detect, which is why fishes kept in an aquarium may ignore people moving just a few centimetres beyond the glass, but dart away in fright if it is tapped sharply.

The prominent lateral line and large eyes of a young Australian bass allow it to hunt in low light conditions.

Other fishes such as catfishes and carp that feed on the bottom in disturbed sediments, or at night, usually have even more specialised sensory organs, including sensitive, fleshy lips, large nostrils which are for smelling and not breathing, or extended whiskery-looking barbels. These recognise potential prey by not only its feel, but also possibly by taste. Fishes that hunt in low light conditions around dawn and dusk usually have large eyes for their size, especially while they are still young and more vulnerable to larger predators, while those that live in the permanent darkness of underground waters (see chapter 11) may lose their eyes altogether, and hunt using a combination of taste and messages from their lateral lines alone.

Berney's catfish showing the nostrils and elongated barbels that convey smell, taste and texture to the fish even in murky waters.

The diversity of native freshwater fishes

Though Australia's freshwater fish fauna is limited in numbers compared to that of other continents, it is surprisingly diverse in terms of the very different families that have moved permanently away from the sea, and in the various ways they have adapted to the challenges of inland living. One of the greatest challenges to fishes living in fresh waters is breeding in environments where silt, fungi, bacteria and sometimes even low oxygen levels are a greater threat to their eggs than in the relatively clear and clean waters of the sea. However, the problem has been solved in many ways, and even in a single family of fishes the range of breeding adaptations can be wide.

Many marine fishes breed repeatedly over their life span, producing large numbers of tiny eggs that hatch into small, feebly drifting or sometimes free-swimming fry. This is an approach that has also been adapted to fresh waters, as many larger inland species are triggered into breeding mode by floods, with their

Young silver perch.

partly developed eggs or milt maturing rapidly with the stimulus of flood. In these species flooding triggers an urge to move upstream as far as possible before spawning, so that their buoyant, fast-developing eggs aren't swept out to sea before hatching.

Grunters are named for the vulgar noises some species make when captured, and this family includes some of the relatively few freshwater fishes that regularly eat plant matter, including fruits and algae. Several grunters have been raised as aquaculture animals, and the silver perch is perhaps the most studied of all native freshwater fish for this reason. Unlike its many more tropical relatives, which mostly lay their eggs in gravels that have been scoured clean by wet season floods, silver perch have moved southwards into the sluggish, often murky waters of the Murray–Darling where they spawn during floods if water temperatures are high enough, producing floating eggs that can hatch in less than 2 days.

Freshwater basses and cods are a single family with diverse approaches to reproduction, with the basses often breeding in estuaries while the larger cods prefer hollow logs. These include the largest and most significant piscine predators, including the Murray cod which can weigh well over a hundred kilograms; the male will guard his eggs for weeks to give them the best chance of hatching. By contrast its close relative the golden perch has taken the same path as silver perch, migrating over great distances upstream to spawn during floods, and also producing large numbers of tiny, floating eggs that hatch quickly. River blackfish and pygmy perch have sometimes been included in this family, but are so distinctive in their own ways that they are looked at in more detail in chapter 12.

Still further south, the short-lived eastern little galaxias (and its two close relatives in Western Australia) also seems to be triggered into breeding mode by floods, but during the cooler months. Often found in the same pools as southern

Southern pygmy perch.

pygmy perch, the orange-striped males and the larger but drabber females of this species are usually fully mature within a year of hatching, and mate as water levels begin to rise. Temperature doesn't seem to have much to do with the urge, as they have been recorded as breeding in over a 20°C range in aquaria, as early as April in wet autumns but sometimes as late as September.

This adaptability is a good strategy for delivering their fry into waters where there are few predators, and an abundance of suitable food. Few larger fishes in the lowland streams of southern Australia are active over winter, and even predators such as mudeyes rarely move into recently flooded areas at these times. Here, within a week or two of flooding, terrestrial grasses have drowned and begin to decay, fertilising a bloom of microorganisms suitable for the tiniest fry. By the time the fast-growing young need larger foods, water fleas and copepods will also have had time to grow and begin to breed, producing an abundant and ongoing food supply.

Many other freshwater and inland fishes (especially small or short-lived species) invest much more of their time and effort in reproduction. Gudgeons and gobies attach their eggs to shaded rocks and other firm surfaces in narrow spaces, where larger fishes can't go. The males not only defend their eggs against smaller predators until they hatch, but also fan them to remove silt and maintain a flow of oxygen-rich water, picking off infertile eggs to prevent the spread of fungi to the healthy eggs. The males may also stay around once the young are hatched, until they are active enough to fend for themselves.

Purple-spotted gudgeon courtship.

Carp gudgeons and their relatives are among the most widespread gudgeons, and include some species that breed in estuaries, but they are particularly abundant in inland rivers so the complex genetic relationships of these species are discussed in chapter 13. Another widely distributed group is the mogurndas, often beautifully marked with purple, red and silver spots which stand out when seen in an aquarium, but which are surprisingly difficult to see in a stream. Some species are found inland including in arid regions, but they are most diverse in tropical streams, and it is likely that several new species will be described as their genetics are unravelled.

Mouth-brooding is the most demanding way to raise offspring, and is practised by several unrelated freshwater groups in Australia. The eggs of mouth-brooding fishes are generally large, and although there may only be a relatively small number of offspring produced each season these are also very much larger than most fish fry. In salmon catfish and mouth almighty it is the males that carry the eggs until they hatch, allowing their partners more time to recover from the massive effort of producing them in the first place. This is the ultimate in parental care, but has its cost as the incubating parent can't feed for several weeks while brooding.

The two saratoga species are tropical survivors from an ancient group of primary freshwater fishes, whose ancestors are not known from marine or estuarine waters (see also the salamanderfish and lungfish in later chapters). Female saratogas not only brood their eggs, but also protect the young after they have hatched, until they become fully independent at around 4 centimetres long. These large-eyed and large-mouthed predators are found in slow-moving creeks and

Seven-spotted archerfish.

reasonably clear backwaters with overhanging vegetation, and can swallow a wide range of prey from frogs to other fishes.

Archerfish are also large-mouthed, surface-feeding tropical carnivores although they are much smaller than saratogas, and not as reliant on animal prey as they will also eat floating fruits and flower buds. They are named for their impressive ability to knock insects down from overhanging vegetation with a jet of water forced through the mouth, and larger fishes may fire accurately to a distance of 3 metres. The primitive archerfish is a freshwater species, but the other two are also frequently found in coastal waters, and one of them is known to breed in both fresh and brackish waters.

Rainbowfishes are a primarily tropical to subtropical family, often found in abundance and sometimes in mixed schools of several species. These are more casual parents than most other freshwater fishes, breeding prolifically over a much longer period of time to make up for their negligent approach. On sunny mornings, males show their boldest and most dramatic colours to attract spawning partners, racing side-by-side at frenetic speeds and glittering like opals or even neon lights as they display to each other.

If a female is suitably impressed the pair may spawn in vegetation near the water's surface, often with other males still trying to get in on the act, and sometimes other females trailing along and eating many of the newly laid eggs. The eggs have threads which tangle into vegetation, partly to keep them out of the reach of bottom-dwelling predators, but also because oxygen levels and temperatures are higher near the surface, so that development is faster.

Crimson-spotted rainbowfish males displaying.

The closely related blue-eyes are smaller than rainbowfishes, and are named for their large, opalescent, pale-blue eyes. Blue-eyes range across a wider range of habitats with one species restricted to a small complex of inland springs, while others are found in estuaries, with the honey blue-eye restricted to the peaty near-coastal waters of the wallum country in northern New South Wales. Blue-eyes may form conspicuous schools during their breeding season, or when they congregate in clear pools. As in the rainbowfishes, males display dramatically, darting with fins flared side-by-side at incredible speeds apparently calculated to drive would-be photographers into a frenzy, but the females are less excitable and lose interest once they have laid several eggs.

Other smaller, schooling fishes may be so abundant that they are among the most important food animals in diverse waters, but are rarely seen unless you set out to net them deliberately. The two species of southern smelt are separated by Bass Strait, one being widespread through the still and slower-flowing waters of south-eastern Australia, and breeding among submerged plants in fresh waters. The Tasmanian species is mainly found in coastal streams, spending part of the year in brackish waters but returning to fresh waters to spawn.

Hardyheads are also small, slender schooling fishes, mostly marine but with a few living permanently in inland waters that may be extremely saline and too hot for most other fishes. Some may have been much more abundant before their habitats were fragmented, reduced and controlled by weirs, but others remain abundant in parts of their range and can be a significant food source for pelicans and cormorants when wet years swell their numbers. Despite their ecological importance and intermittent abundance, little is known of the breeding biology of many species.

Flyspecked hardyheads.

Glassfishes are mostly small tropical fishes which are most active at night; the olive perchlet (see chapter 10) was once abundant all the way down the Darling River, into the warmer parts of Victoria and southern South Australia. Although some are almost as transparent as their name suggests, with their gut and bone structure being clearly visible, others may have strongly marked patterns or scales, and the giant glassfish is so large and strongly patterned it could pass as a coral reef predator. The smaller species often school in clear, slow-flowing or still waters in huge numbers, living and breeding among submerged plants, and only the rainbowfishes are likely to be more abundant – or more useful as a prey animal.

Eel-tailed catfishes are a very distinctive group with a fringe of fin around most of the rear part of the body, some of them looking like a deep-bodied type of eel apart from the thick, fleshy barbels (whiskers) around their mouths. Most of this family are tropical to subtropical, including some boldly striped marine and estuarine species, but others are found across a wide range of inland habitats and even into desert regions. Two closely related species are also found in southern Australia: the freshwater cobbler in the south-west, and the freshwater catfish through much of the Murray–Darling as well as closer to the central-eastern coast.

Catfish barbels are sensitive organs of both taste and touch, well adapted to foraging for insects, crustaceans and molluscs, and some species will even crawl around in shallow waters when feeding or when the urge to migrate upstream through floodwaters comes upon them. Little is known about breeding in this family, apart from the two more southern species; both are good-quality edible fishes so they have been considered as potential candidates for aquaculture. The males in these species build a nest of pebbles or smooth gravel in which the eggs

Black catfish – an eel-tailed species.

are laid, raising them above any surrounding silt, but this may be abandoned if water levels drop too quickly before hatching. Other species that breed in sandy or stony places don't build a nest in the few cases where anything detailed is known of the breeding behaviour.

Fork-tailed catfishes comprise a separate family that remains more closely tied to its marine origins, most of them marine and estuarine species with only a few found in fresh waters. As already noted above these are mouth-brooders, and in the few cases where breeding localities are known even the inland species seem to favour coastal and near-coastal localities. These thick-bodied beasts remind me of a chunky shark, and with their large mouths have a similarly voracious approach to their diet, even feeding on fallen fruit when nothing else is available. Fork-tailed catfishes are usually solitary when adult, but they tend to form loose schools while still fairly young.

Human impacts: introduced fishes

Introduced fishes have changed the nature of many wetland habitats worldwide, and threaten the future survival of many native fishes and frogs. Some were deliberately introduced as game fishes, and others were careless releases by aquarists. The prolifically breeding gambusia or plague minnow was released because it supposedly did a particularly good job of mosquito control. Now one of the most widespread vermin fishes in the world, this livebearer swarms in warm, shallow water across most of temperate to subtropical Australia. It has even been introduced into isolated desert springs where there were no mosquitoes, threatening the survival of specialised native fishes instead.

In the cooler streams and rivers of the south, three species of trout introduced 150 years ago as game fishes have had dramatic impacts on populations of many smaller native fishes, and it is possible that they have exterminated some galaxiids

Plague minnows.

before humans even had a chance to become aware of their existence. In the lower-lying streams of southern Australia huge carp stir up bottom sediments as they feed, so that plant beds which were once a significant breeding and shelter habitat for smaller fishes have disappeared in many places. Tilapia have been temporarily controlled in more northern waters, though at considerable expense, and if these were left to spread they would become yet another problem for native species just as they already are in many tropical countries. The dismal list continues with many other vermin species from redfin perch to weather loach, goldfish and barbs, guppies and various other livebearers already established as localised populations, many of these dumped into rivers by careless aquarists.

A brown trout hovers in a shallow stream.

5

Frogs, reptiles and mammals

Reptiles and amphibians are closely linked in many people's minds, though the physical differences between them are considerable, and they have very different ecological niches. These can be generalised simply: the soft, scale-less skin of amphibians means that they have only a limited tolerance for drying out, and their even softer eggs must be laid in water or in reliably wet places. The only native amphibians in Australia are all frogs, and because of their specialised needs most of these are wetland animals for at least a part of their life cycles.

Their sensitive skin is probably a significant part of the reason for the dramatic decline or even disappearance of many amphibian species worldwide, including in Australia. Whether they are unable to cope with new strains of chytrid fungi (the major suspect in the disappearance of many frogs worldwide), increased ultraviolet radiation, an increasing range of agricultural toxins, or other less obvious changes over the past few decades is unclear, but there is little doubt that their decline is linked to human impacts on the wider environment.

By contrast, the scaly skin of a reptile is designed to keep moisture in so desiccation is not a problem except in extreme conditions, and many reptiles live in arid places. Others have returned to the waters to hunt or fish, though these are usually also able to live in drier conditions for much of their life cycles. Although the leathery shell of a reptile egg means that they don't need water to breed, it also means their eggs would drown if actually laid *in* water, and to get around this problem some aquatic reptiles have become livebearers whose eggs develop into miniatures of the adult within the mother.

The leathery and impervious head of a saltwater crocodile.

The life cycles of frogs

For all the apparent delicacy of frogs, their need for water and the vulnerability of their young to predators, their ecological role in wetlands is far more significant in most places than that of reptiles or mammals. And as would be expected for a group of animals with a long evolutionary history on this continent, native frogs occupy many habitats and use them in many ways, at the same time as providing a major food resource for a varied range of predators from water tigers to water bugs, fishes, mammals and birds.

The young or tadpole stage of frogs is very different from the more specialised adults, with a long gut that makes it possible for them to digest plant matter. This is coiled up tidily in their bulging bellies to make them more efficient feeding machines: a larval stage designed to put on weight as quickly as possible, so they can turn into adult frogs as quickly as possible before the waters they live in dry out. Their lips are designed to scrape and break up algae and other soft foods, although some may also scavenge or be partly carnivorous.

At metamorphosis, the tadpole tail begins to be absorbed at more or less the same time as the small rear legs first appear, so that the young frog is increasingly

Brown treefrog tadpoles and developing froglings.

able to move around on its limbs as the tail shrinks and becomes less of a burden. By the time the tail has become small enough to not drag behind, the frogling can leave the water for long periods of time. Adult frogs feed primarily out of water, and the most common diet is insects in one form or another. Sometimes this may be a specific group such as termites in the case of some desert frogs, but more usually a wider range of species is taken, basically whatever can be readily caught in season.

Smaller frogs feed mostly on smaller insects, while those with larger mouths are more opportunistic and may even take young snakes and rats, or hard-shelled prey such as snails. Apart from a food supply, the main requirement for their survival is shelter, as many other wetland animals prey on adult frogs if given the chance, including humans in many parts of the world. Some prefer vegetation such as tussocky plants to shelter among, while others may hide under rocks or fallen timber, which is also an excellent insulator from solar heat.

Male frogs call to attract a mate, and the calls of most species are so easily recognised once you are familiar with them that they are a good way to identify species without having to spend all night trying to catch a solitary male in a dark swamp. However, the calls may slow down so dramatically on a colder day that you may imagine you are hearing something new and unfamiliar. Depending on the species, frogs may call from among dense vegetation, in smaller pools of water near larger wetlands, or while floating in open water. Many frogs call in groups, setting up a chorus triggered by the first in each group to call. This may be to distract females from the male that makes the first call, but probably also helps to confuse potential predators.

Wetland frogs

Apart from a small number of tropical frogs which are mostly semi-terrestrial, the great majority of native frogs are classified into just two major families: the southern frogs, a diverse family that has evolved and diversified on this continent, and the treefrogs which moved south from Asia a very long time ago. Most climbing frogs in Australia are treefrogs, with their greatest diversity towards the tropics though there are also three species as far south as Tasmania. Nearly all of these are active jumpers as well as climbers, and are usually found near streams and other more permanent waters. Some such as the brown treefrog (see chapter 8) will travel long distances overland and can appear in some surprising places, while the common green treefrog is a familiar resident of less than pristine habitats such as toilet bowls in warmer climates.

Southern frogs were already diversifying in Australia long before the treefrogs arrived. Many of the smaller species (known as froglets) are semi-terrestrial, congregating in large breeding groups soon after rain fills ephemeral pools and grassy depressions. The variable patterning of many froglets can make it difficult to tell non-calling males (and also often females) apart, though you can usually work backwards from the geographic area where a froglet is found to work out what species you are most likely to be holding in your hand.

Despite their common name toadlets are true frogs, breeding in marshy patches among forest or rocks, while the seriously endangered corroboree frog is found above the tree line in wet, grassy areas and sphagnum bogs. Southern toadlet groups include many species with striking colouration that presumably warns predators they are toxic, but northern toadlets are such small and cryptic animals that they don't have bona fide common names, just awkwardly contrived ones bestowed on them in recent years.

Burrowing frogs are a diverse group of animals which aren't necessarily closely related, mostly coming to the surface to feed and breed after periods of rain. Most of these are large, blunt-featured creatures, often with cat-like eyes and broad feet with little sign of webbing, better adapted to burrowing than to swimming. Spade-foot 'toads' (so-called because of their warty skin) are among the most specialised burrowers, with shovel-like feet and almost spherical bodies that hold water during periods of drought, allowing them to surface only after the occasional rainfall when they breed in temporary pools.

Marsh and banjo frogs are as diverse as the habitats they are found in, including some species that are adapted to arid areas and can burrow moderately well,

Striped marsh frog.

though most are found in more permanently wet places. These are generally medium-sized to fairly large frogs, which lay their eggs in a conspicuous, floating foam mass that probably provides some protection against predators, and their mating calls are usually very distinctive as well as loud.

In my local area the distinctive 'tock … tock … tock' call of striped marsh frogs is reminiscent of a leisurely game of tennis played on a grass court, but the manic roar of a crowd of pobblebonk males in full croak is even more entertaining, each one shouting 'bonk' as quickly as possible after any other male dares to call, to create the garbled 'pobblebonk' sound that gives them their name. In a good breeding year pobblebonks can be so loud you have to shout above them if you have anything to say near our swimming dam, and newly arrived visitors have assumed that for some bizarre reason we are playing the commentary of a race meeting at full volume!

Freshwater turtles

Freshwater turtles were long known in Australia as tortoises, a word better confined to the high-domed terrestrial animals with very different habits found in

Macquarie turtle.

other warmer regions of the world. The word 'terrapin' has also been used, referring to more streamlined, semi-aquatic tortoises, but freshwater turtle is now widely accepted as appropriate for all Australian species. The tropical pig-nosed turtle is the only species that has flippers like those of marine turtles, and is discussed in chapter 17.

All other freshwater turtles have legs and some can travel considerable distances, giving them the option of abandoning places where water quality is deteriorating, or where there is little in the way of food. Others prefer to burrow into mud and aestivate through hotter, drier periods, even if these may last for months, or may just bury themselves under fallen timber where temperatures aren't likely to climb too high. In the cooler parts of eastern Australia most freshwater turtles hibernate during the cooler months when they are less active, some emerging from water to do so, while others stay underwater for days or even weeks at a time.

There are two main groups of freshwater turtles: the long-necked (snakeneck) types, and short-necks. Long-necked turtles are found from southern Victoria and south-western Australia northwards along the coastal belts up into the tropics and New Guinea, as well as throughout the Murray–Darling. Short-necked turtles are found inland in the Cooper's Creek and Murray–Darling drainages, but are most diverse in eastern and northern Australia, where it seems almost every larger river system has its own distinct form.

The short-necked group known as saw-shelled or snapping turtles is largely tropical in range, with a few distinct forms found as far south as New South Wales. The strong jaws of these animals allow larger animals to feed on mussels, either by crushing the shell or by biting through where the two valves close together. There are also several oddball turtle groups which are mostly river animals as well as the unique western swamp turtle, now only found in two ephemeral swamps in south-western Australia where it is surrounded by fox-proof fencing.

Saw-shelled turtle.

Nearly all freshwater turtles prefer to feed underwater, hunting smaller fishes, tadpoles, shrimps and crayfishes, and they will even feed on carrion in some cases. Some individuals within the same population may have their own specialist tastes, feeding selectively on particular animals, or they may just be better at catching them compared to their relatives. They will also dive to avoid predators, and a disturbed animal may remain submerged for surprisingly long periods of time.

Larger turtles were a favoured Aboriginal food wherever they were found, and still are in some places. Crocodiles are the only natural predator able to eat whole adult turtles, though larger salmon catfishes will swallow hatchlings and even young adults, and eels sometimes feed on them in southern waters. All native species have scent glands which appear to be used for defence, producing a vaguely offensive odour if they are disturbed or frightened, and some populations of snakeneck turtles are even known as 'Murray stinkers' for obvious reasons if you have ever carried one around in a pocket as I have.

Breeding females will usually seek out a nesting area well above potential flood levels, travelling considerable distances and even climbing well above the normal water level of the rivers they live in if necessary. The eggs take months to hatch, and in some cases live young have been found in them nearly 2 years after the nest was dug. The nest is vulnerable to predators through all this time, originally native animals such as goannas but now mostly foxes. Young turtles are also vulnerable as they leave their nest and scramble their way to the water, but once submerged they are among the most elusive aquatic animals and it is rare to see or catch one.

Lizards, snakes and crocodiles

While turtles are relatively sluggish animals that can be clumsy out of water, relying on their shells for partial protection from other animals, most other aquatic reptiles are more streamlined and can be surprisingly fast-moving, both in water and on land. Crocodiles are probably the best-known aquatic reptiles due to the media getting excited about the occasional human they eat, and are discussed in chapter 10.

Lizards are primarily terrestrial animals, though quite a few species of skinks live in the vicinity of streams and wetlands, and some of these are active swimmers which will also feed on small fishes trapped in shallow pools. Although mostly found around creeks and wetlands and excellent swimmers which take to the water readily, they prefer to lay their eggs on higher and drier ground wherever possible.

Among the so-called dragons, only the eastern water dragon (see chapter 12) is truly aquatic, but in tropical Australia several of the smaller monitor lizards (goannas) are found in and around streams or among mangroves in tidal areas. These feed above water on anything from insects to frogs while they are still young, graduating to larger prey including fishes and even carrion once they are large enough. One species seems threatened by the spreading range of the cane toad: Merten's water monitor can swallow toads large enough to kill it, and anecdotal evidence suggests dramatic population declines as the cane toad invasion sweeps westward across the tropical north.

Few Australian snakes are truly aquatic, but most can swim well and quite a few include frogs as a major part of their diet, so they are often found near water. In southern Australia visitors of this kind are mostly venomous, but the dangers tiger snakes, brown snakes and black snakes pose are greatly exaggerated. The

Merten's water monitor.

relatively small-headed and usually very shy copperheads are active in much colder weather than other temperate climate snakes, and I have watched them swimming and diving in pursuit of prey on sunlit days at the tail end of winter, emerging to warm up in the sun every few minutes.

The more venomous snakes are replaced by a very different range of semi-aquatic species in the tropical north, from keelbacks to various mangrove snakes that hunt for fishes trapped in tidal pools. The most aquatic of these are the two species of file snake (see chapter 14), which share their tropical habitats with the water python, a more versatile species that feeds opportunistically on a wide range of foods both in and out of water, and when wetlands are dry it will even forage for rats in deep cracks in the ground.

Human impacts: alien amphibians ... and a reptile

Originally introduced to control insect pests in sugarcane crops, the cane toad prefers to eat almost anything else instead, and both tadpoles and the adults are also highly toxic to indigenous predators that mistake them for the many native

Cane toad.

frogs they can eat without harm. Cane toads continue to spread westwards across the far north, and if some of the theoretical models of their potential spread are correct, they are also likely to become one of the dominant wetland vermin species much further south as global warming progresses.

But even indigenous frogs are sometimes translocated to places where they don't belong. While the common green treefrog often appears in southern cities among bunches of bananas it doesn't seem to thrive in these alien environments, but the eastern dwarf treefrog has already become a breeding resident in parts of Melbourne, well south of its natural range.

The only feral wetland reptile in Australia is the red-eared slider, a gaudily striped central American turtle which was smuggled in as it was a popular pet species in many other parts of the world. Occasional animals are caught or reported around southern Australia, and though these have shown no definite sign of successful breeding as yet, the species is on the World Conservation Union's list of 100 most invasive species, so any seen should be reported.

Red-eared sliders.

Mammals

It may seem bizarre including mammals in a chapter mainly devoted to amphibians and reptiles, yet the only two truly aquatic native mammals are mere bit players in the greater scheme of things, sharing many habitats (and sometimes even the same tunnels) in their roles as minor wetland predators. Tied closely to relatively short sections of streams, their impacts on the animals which live below the water's surface are not great compared to predatory waterbirds, though they may have had a greater ecological impact before they were hunted for fur on a large scale after European settlement.

Water rats and platypuses usually don't travel great distances, foraging instead within a well-defined home range. In turn that means their populations are regulated by the food resources available throughout the year rather than seasonal abundances, so family groups are the largest gatherings likely to be found together and then only during the breeding season. That doesn't make mammals of less significance than birds, though it is their conservation that is of more concern than any strong ecological influence they bring to bear, as they can easily be driven from a section of river or wetland, never to return.

The platypus is among the most fascinating and reptilian of all mammals, but its history and evolutionary origins have been repeated so often that there should be no reason to detail these here, and it even has a regularly revised book (Grant 2007) dedicated to it alone. Feeding mostly around dawn and dusk its main diet is a diverse array of invertebrates up to the size of a small freshwater crayfish, though it will also take tadpoles and even small fishes if given the opportunity. This is a truly unique and iconic animal, and people come from all over the world for a glimpse of one in the wild.

I have guided tour groups to watch platypuses on a nearby lake and the experience can be magical, walking in before first light among constellations of

Platypus.

glow-worms beneath the tree ferns, literally as bright as the stars above. As the light brightens and the mist over the lake begins to lift the canoes are roped together so that the guide can do the manoeuvring, as the paddle must be feathered so it makes no sound. The first person to sight a platypus whispers its direction … two o'clock, and there it is. A miraculous little animal drifts as it grinds up a caddisfly larva, then dives to see what else there is for breakfast on this fine morning, as its species has been doing for many millions of years.

The water rat is far more adaptable yet less glamorous despite attempts to rename it with various attractive but localised Aboriginal words, such as rakali. Almost otter-like compared to other native rats, with a tidy toothbrush of bristles along each side of the nose, it is found from subalpine areas to the sea. When the water rat family that lived around a suburban lake where I used to catch yabbies as a child moved away as their prey declined in numbers, I was pleased to find them re-established among rock piles under a jetty in nearby Port Phillip Bay.

Although water rats are much more common than the platypuses they are seen less often, the main sign of their presence usually being middens made up of the remains of meals on rocks and logs near the water. An adaptable carnivore, the water rat will take crayfishes, mussels, surprisingly large fishes if they can be caught in a shallow pool, and even whole lamb chops intended to be cut up for yabby bait. Another dawn and dusk feeder, like the platypus it also retreats into tunnels it has made in inconspicuous places along the shoreline by day.

Origins: Gondwanaland and Asia

Australia was once joined to several other landmasses including parts of South America and Antarctica, as part of Gondwanaland. As these pieces drifted apart, each carried related animals and plants which have now had a hundred million years and more to evolve in different directions. Indigenous groups with a Gondwanaland origin range from eucalypts to Queensland lungfish, banksias and most native freshwater turtles, southern beeches and platypuses, and many of these still have close relatives on other former pieces of Gondwanaland.

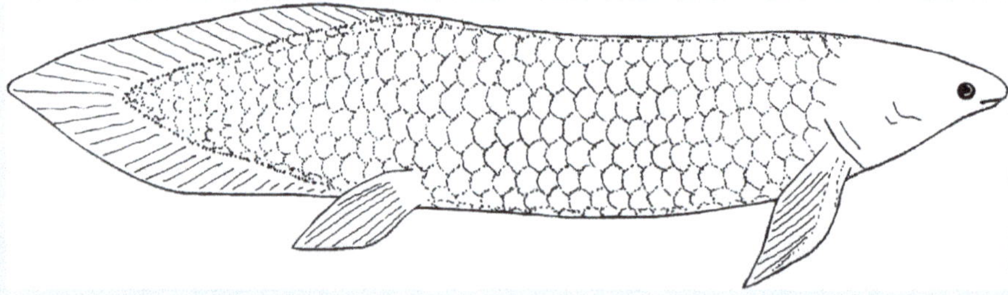

Queensland lungfish.

As Australia continues to creep northwards towards Asia at far less than a snail's pace, pushing up the mountains of New Guinea as it goes, an ever-increasing variety of Eurasian animals and plants has been arriving on our shores. This process has been going on for millions of years, and many of these newcomers have had plenty of time to adjust to Australian conditions, ultimately evolving into new and distinctive species here.

Among the reptiles and amphibians relatively recent arrivals include water monitors, some of which are also found up into Asia. The many native treefrogs have been here for so long they have developed into a distinct subfamily, different enough from overseas representatives of the group that it has even been suggested they should be separated as a family of their own. All native rats were also originally from Eurasia though many of the Australian species are now very different from their northern relatives, including the highly specialised water rat, a beautifully adapted, somewhat otter-like species that now shares many riverine and lake habitats with the platypus.

Green treefrog.

6
Waterbirds

Waterbirds are the most conspicuous grace notes in wild wetlands, and even the most unappealing urban wetland can be made more interesting by a pair of ducks, or a passing heron. As intelligent, far-ranging creatures finely attuned to the vagaries of Australian climates they can give us an idea of the depth and water conditions in a wetland, acting as guides and indicators to the types of animals that live below the surface, and are also acute weather prophets able to sense changes taking place thousands of kilometres away.

With their power of flight waterbirds can carry minute animals, spores and waterplant seeds between wetlands, like the proverbial stork of European mythology delivering complete biological communities to newly flooded wetlands. And those powers should not be underestimated; over the past century the cattle egret has extended its range over much of the world, and even such apparently clumsy flyers as swamphens with their gawky, trailing legs and awkward vertical take-off have reached New Zealand many times under their own steam, and have been crossing the Tasman Sea for so long that some of their earlier relatives have had time to develop into distinct, flightless species there.

The migratory urge

Many wetland birds such as kingfishers, whistling ducks and darters are widely distributed across Australia, yet are far more abundant in the north. In some cases such as magpie geese and brolgas this is the result of shooting or drastically reduced habitat as these birds were once more widespread and abundant, and given the slightest encouragement they will return. Others wander more widely although

Great cormorants rest on a mudflat after fishing.

unpredictably, following the rains wherever these may fall, or travel vast distances to the Northern Hemisphere and back along well established pathways.

Although insects also have the ability to fly between wetlands, and some may abandon the water in large numbers at the same time, there is little evidence of deliberate organisation or recognisable patterns in their group flights. By contrast, the behaviour of many birds is so precisely coordinated and goal-directed that there has never been any doubt that they deliberately migrate in search of breeding grounds and feeding areas. What is still in doubt even now, after decades of intensifying research and understanding of their navigation methods, is just how they find their way.

Among wetland birds the regular migrations of shorebirds over tens of thousands of kilometres are probably the best documented, with many species following known routes to Northern Hemisphere breeding grounds while food supplies are abundant there, and returning to Australia (or other countries as appropriate) for a prolonged feeding break during which they moult and grow a set of new feathers for the next lap of their journeys.

The journey north is tightly choreographed as the departing breeders have only a short window of opportunity in places with a brief summer, such as the Arctic tundra. Returning birds have less pressing business, so they may reappear more

Sharp-tailed sandpipers with red-kneed dotterels.

sporadically in their Australian haunts, and over a longer period of time. Shortly before a northwards migration begins the birds become increasingly restless, gathering in vocal groups and indulging in odd humours such as restless jumping or singing a specific song not heard at other times, or making noisy practice flights that fizzle out for lack of enthusiasm after a few minutes.

But when the time is right, often towards dusk, the birds form up into the vee-formations and lines which minimise air turbulence for those following the leading birds, and begin their epic journeys. It is probable that the lead bird drops back after a time so the next in line takes over this more laborious role, before falling back itself in time, but as they are well out at sea and kilometres up in the air by this time there are no human observers around to verify this.

The most likely suggestions for the ways in which migrating birds guide themselves include by the stars and by the angle of the sun, which changes appreciably as they move hundreds of kilometres northwards each day. Overcast conditions are no problem, as birds can see polarised light so the precise position of the sun is no mystery to them. As they get closer to their goal, birds that have flown this way before fine-tune their course using familiar landmarks, an ability younger birds must be able to pick up rapidly, as on their next journey they will not necessarily be with experienced travellers.

Since the discovery in the 1960s that birds can detect magnetic fields, this has become the hot contender as the number one navigational aid for migrating birds.

Quoting from *Shorebirds of Australia* (Geering *et al.* 2008), 'the magnetic field takes priority if the different [navigational] methods do not agree', yet none of the books I have read on this subject seem to discuss a geological phenomenon that throws a big, fat spanner into the works. Gradual changes in the Earth's magnetic field could obviously be accommodated for as each new generation of birds learns to make minor adjustments to a traditional flight path, yet every few hundreds of thousands of years the Earth's magnetic field does a complete backflip, known as a magnetic reversal.

After a shift of this kind do entire species of birds fly off into the Pacific or into the Antarctic winter, never to be seen again? Somehow I doubt it, because it is also known that they can make many other kinds of adjustments to their flight path on their way. Migrating birds must be able to allow for unseasonal winds, possibly even factoring in storms they may be able to detect from a thousand kilometres away or more through deep, powerful infrasound waves. Or is their ability to detect magnetic fields much more subtle that the simple north–south notion humans seem obsessed with? Perhaps they actually follow the long-embedded magnetic reversals in the rocks that have formed over millions of years deep under the sea, as has been suggested for sharks that cross oceanic distances on their regular migrations.

They have been doing this very successfully for millions of years, which means that provision for dealing with magnetic reversals must be built into their genes, and the greatest threat to their long-term survival is the loss of many of their resting habitats along the flightways. The international Ramsar treaty has had considerable success in coordinating conservation efforts among the many nations involved, but there doesn't seem to be any way of stopping rampant development in some key habitats and places.

Human impacts: waterbirds, environmental flows and agriculture

Intelligent use of water became an increasingly sensitive issue over the course of the prolonged drought which opened the 21st century, and the water rights which had been liberally handed out in earlier years were finally recognised as unsustainable. Yet some forms of agriculture in arid areas continued to claim the lion's share of river flows, with one cotton farm continuing to fill its huge storage dam from the Darling River from a quota greater than all of the environmental flows allocated to the entire Murray–Darling Basin. These are the *theoretical* minimal amount of water needed to keep wetlands alive, but in drought years they have rarely been honoured.

Environmental groups justifiably blame irrigation farming as the primary reason for the ever-deteriorating condition of inland rivers, but an ever-increasing proportion of water rights are also used to water lawns and fill 'ornamental' ponds

This large pond in inland Victoria is kept filled purely for so-called aesthetic reasons, and is effectively an evaporation basin that wastes around 50 million litres of water each year.

which are really just evaporation basins in these arid areas. Rice cultivation has often been targeted for criticism even though it doesn't use much more water than other forms of irrigation farming, yet there is no reason even this couldn't be grown in inland regions of Australia during years of abundant rainfall, and these peak-year crops can be stored as live, dry seed for at least 5 years.

And rice fields don't benefit just humans; I have seen acreages where widely spaced brolgas stood apparently lost in thought under a bright moon, while in Italy it is estimated that a quarter of the surviving bitterns of that country live permanently in the risotto fields of the north. Other birds such as magpie geese have been a problem for good rice yields in northern Australia and were the cause of the demise of the fledgling industry there, but planting upland rice (which doesn't need irrigation) looks to be a promising method of minimising conflict between growers and wildlife.

Other migratory wetland birds live out their entire lives within Australia, following rain and other seasonal changes, and breeding as suitable conditions become available. Even some indigenous shorebirds have adopted this apparently random behaviour pattern, although their close relatives continue to commute from the Southern to the Northern Hemisphere. Endemic ducks, geese and swans include some of the best examples of waterbirds which seem aware of climate events which take place thousands of kilometres away, but apart from knowing what it is that draws them we still have little understanding of how they know where to fly, or when.

Ducks that move on in droves when rain is falling heavily elsewhere (see chapter 9) may have a good idea of where the rain is pouring down from shifting isobars, but not necessarily how far away the action is. My suspicion is that they aren't such brilliant weather prophets as folklore often suggests; for example, swans that 'always' build their nests just above the likely high water level in any particular winter are almost always flooded out during wetter years. Yet species such as pink-eared and freckled ducks breed primarily in ephemeral wetlands such as lignum swamps, where explosions of planktonic life provide immense volumes of food, so they must somehow be able to locate such newly flooded habitats while these are at their most productive.

Eggs, nests and later problems

Birds have inherited a relatively hard-shelled egg from their reptilian ancestors, but with their overall evolutionary trend towards reduced weight and flight efficiency, not one of them has been able to develop live-bearing habits as an alternative because the additional weight involved would be a crippling burden for the females. Instead, each individual egg is formed as quickly as possible, and is laid in a nest where one or both partners can keep the completed batch warm until they hatch. While this allows both parents the freedom to fly and feed for as long as possible, it does create other problems, which for waterbirds include keeping the shelled eggs out of water as they would drown quickly if submerged.

Many waterbirds nest close to where they feed, which is not difficult for egrets, spoonbills and others that join in loose colonies in seasonally flooded shrubs and trees such as paperbarks. Others must raise their nests above the waters with humbler materials, building mounds of vegetation and sometimes mud from various sedges and grassy plants such as water-ribbons. A few less aquatic wetland birds such as reed-warblers weave their nests well up among reeds, the cups tied around three stiff stems for additional stability, but still low enough that the adults are safe from attack by swamp harriers and other small raptors.

Although apparently free-as-a-bird to come and go once their offspring have hatched, the parents must then keep up the food supply to raise them to independence as quickly as possible. The impressive growth rate of young birds comes with an even more impressive feed bill, because even really large birds such as brolgas *must* get their chick to flying-size within around 3 months. In turn, that means the adults should be as close to their food sources as possible, which for many waterbirds also ties them to the period of greatest abundance that starts not long after wetlands refill, whether during the wet in the north or spring in the south.

And then comes a tragic time: persuading the spoilt young, already as large as and sometimes heavier than their parents, that it is time for them to find their own

Dusky moorhens feed their demanding young.

food. Every birdwatcher has watched frustrated parent birds trying to drive their insistent offspring away, and this section of the book was finished within 2 hours of watching a white-necked heron attacking its chick on the wall of our swimming dam. As far as the pursuit of natural history goes it was the ideal situation, with both Jan and I sitting indoors in our favourite chairs and not even needing binoculars. The bedraggled adult, its neck mud-stained from months of foraging to raise the chick, finally resorted to dive-bombing the snowy-white and bewildered youngster, so that it literally dropped flat to the ground time and again.

Whenever the parent retired to a nearby eucalypt branch the youngster would try to join it again, and the two would fall out of the tree as they squabbled, beginning the cycle again. The entire episode ended amusingly when grey kangaroos that use the wall as a pathway arrived, scaring the young heron away, in turn scaring a startled female kangaroo into a spectacular sideways leap that ended in the dam. The kangaroo was grazing on the wall minutes later, but it was perhaps 15 minutes before the adult heron returned, stood looking around for another minute and, apparently satisfied that it had shaken its pursuer, moved on at a more leisurely pace.

The diversity of wetland birds

Most major groups of birds include species that live either seasonally or permanently in and around wetlands, including more specialised members of essentially

Nest of the Australian reed-warbler.

terrestrial groups such as honeyeaters, robins, whistlers and even some smaller parrots. Songbirds such as the cisticolas with their curious, insect-like buzzings and wheezings are rarely found far from water, and Australian reed-warblers are usually seen only as a flash of grey among the reed beds where they live and breed, and where their rich, varied and melodic calls are almost the defining sound of such habitats across much of Australia.

Other wetland birds are more obviously adapted to water, with ducks, geese and swans probably more strongly associated with wetlands in most people's minds than any others. Like sheep and goats, ducks and geese are not always easy to tell apart, and some of the birds we usually call ducks are really miniature geese in terms of their behaviour and ecology, living near water but grazing nearby. Of the true geese the magpie goose is now usually regarded as a northern species although it was also abundant in southern Australia until it was hunted and poisoned out, or abandoned former habitats which had been drained; it is now reappearing in parts of its southern range in small numbers.

Shelducks are the largest of the goose-like ducks and may spend much of their time far from water, although they also feed on various types of algae and even floating fairy fern. The Australian shelduck of the south-east and south-west aggregates in large flocks after the breeding season, usually on the fringes of large bodies of water where they can avoid predators while their new feathers grow out. The other native species is the radjah shelduck, normally tropical although a

Australian shelducks.

vagrant pair took up residence for a few months along the river near my home, inadvertently creating a nuisance on weekends when busloads of twitchers from Melbourne would crowd the gravel road near the flats where they fed.

Whistling ducks are also goose-like and are named for the sound of their flight, forming huge flocks in northern parts of Australia, but outside their breeding season they also make unpredictable sorties much further south on the mainland. Like the wood ducks which are common in temperate Australia they often graze long distances from water, but wood ducks are a more family oriented bird which forms small flocks even around farm dams if a suitable hollow tree is available for breeding.

Dabbling ducks don't dive, feeding instead by browsing in shallow water and sieving water through their beaks while standing on their heads to reach the bottom. Black ducks will appear almost anywhere there is water, and will even nest in long grass at the side of an apparently barren farm dam. Teal are a little fussier about their nesting sites, although chestnut teal will lay eggs in a range of places from islands and tree hollows, to places well away from the water's edge. Females of most dabbling ducks raise their young without help from the males, except in the case of the chestnut teal, which seems to form a more lasting pair bond.

Musk ducks are fascinating, largely aquatic birds with their legs set well back on the body, swimming with their bodies sunk deep in the water, and rarely flying in daylight. Strong divers that can stay underwater for a minute or more, they feed mostly on insects but can also catch crayfishes, mussels, and sometimes even frogs. Courting males use their powerful legs to kick up a geyser of water that can be seen from a long distance, but leave the females to raise their young alone in a reed nest, although the females may also lay eggs cuckoo-fashion in the nests of other ducks.

A pair of chestnut teal.

Black swans are one of the native animals that were so different from their equivalents in the Old World that they helped create the image of the Antipodes as a place set apart, and when the first reports of these reached Europe where all

Male musk duck.

swans are white, they were greeted with some incredulity. Graceful when swimming, swan take-off involves a laborious beating of wings, but once in flight a flock is a beautiful sight, the regular rush of their wing beats blending with melodic, bugling calls to create one of the most characteristic and haunting sounds of open water habitats.

The smaller members of the rail family are cryptic birds that hide within vegetation, and are usually only briefly seen when they dash across an open area. Crakes are particularly small and extremely shy birds with blunt beaks, while rails are larger with more elongated beaks. The larger rail relatives are much less elusive, with purple swamphens commonly grazing across suburban lawns near wetlands, and Eurasian coots sometimes forming flocks of hundreds of birds that feed across the open waters of shallow lakes. While the smaller rails and crakes feed on invertebrates as well as some seeds and plant matter, their larger relatives are more markedly herbivorous, and I have seen a small group of swamphens feeding high up on the fruits of a prickly currant bush, many metres above water.

All members of the rail family have very long and elegant toes that spread their weight out on semi-liquid mud, also allowing the relatively lightweight dusky moorhen to walk across waterlily leaves. At the very lowest end of the size scale,

Baillon's crake running on azolla.

Baillon's crake looks like a quail with oversized chicken legs grafted on, and can run even across mats of tiny, free-floating azolla ferns as long as it doesn't slow down. Watching closely, it can be seen that however fast it moves, the leg behind has sunk below the water's surface, even as the forward leg flattens over the next patch of ferns.

Other waterbirds are more strictly carnivorous, with a wide range of strategies for capturing their prey. Long-legged, stalking species need reasonably open waters to hunt successfully, although they will also feed in weed-choked waters if enough food is present to justify the extra effort. The down-curved bills of ibises are used as a probe, locating smaller animals such as insect larvae under water, and they will also forage across open paddocks during wetter periods when grubs and worms must rise to the surface to avoid drowning.

Spoonbills are related to ibises but their spoon-tipped bills are designed to sieve out smaller prey from small fishes to insects and crustaceans, using a side-to-side motion of the beak, and they must feed for long periods of time to satisfy their needs unless food is abundant. The beak reflexively snaps shut on anything substantial that comes in contact with the inside of the spoon, and as the prey is

Royal spoonbill.

gulped down water pours inelegantly from the mouth. Spoonbills usually travel in pairs and small groups, but are similar enough to ibises in many other ways that they may nest together in mixed colonies, often in flooded vegetation such as sedges or shrubs.

Herons and egrets vary considerably in size and appearance but are related, the egrets being most conspicuous as they are white, with some developing striking plumes during their breeding season. All species feed on fishes, frogs, larger invertebrates and sometimes on young birds and small mammals, foraging by preference in relatively open and gently sloping areas. While herons nest in smaller groups or as pairs, egrets prefer to make up mixed colonies where they may be joined by spoonbills and ibises. The smaller herons are cryptic and not often seen unless you are specially looking for them, including the mostly nocturnal nankeen night heron, and several species of bittern which are so well camouflaged and secretive that it isn't even certain how common some of them are.

Herons and egrets are mostly elegant fliers that travel with their heads tucked back into the body, and legs trailing behind, so it isn't hard to tell them from storks and cranes which fly with necks outstretched. The jabiru is the only native stork, a clumsy hunter that sweeps its beak through shallow waters to seize fish, snakes, frogs and any other larger animals it bumps up against. Brolgas are the most widespread of Australia's two native cranes and, unlike herons, egrets and storks include a considerable amount of vegetable matter in their diet. In the north this is often the tubers of bulgurru sedge, while in southern Australia they will forage in stubble fields for fallen grain, crickets and even mice.

The most specialised piscivores dive for their prey, and cormorants are particularly abundant and widespread in a wide range of habitats from ephemeral waters to farm dams, estuaries and open sea bays. Highly effective predators, these are

A small flock of cattle egrets watch a distant swamp harrier.

Australasian grebe.

discussed along with the more elegant and unusual Australasian darter in chapter 10. Grebes are still more aquatic, to the point where most species don't make a habit of flying, and with several specialist feeding techniques and habits these also make their bows in later chapters.

Kingfishers are ambush predators found in a wide range of habitats where the water is reasonably still and clear, including estuaries, mangroves and larger streams. From their favoured perches they dive at startling speeds to take fishes, frogs, crayfishes and crabs, killing the victim by beating it against a branch rather than with a stab wound. The nest building techniques of kingfishers are as specialised as their hunting techniques, with a tunnel being made into a creek bank or a termite nest.

The several raptors that are most commonly associated with wetlands are never abundant, although the mostly coastal and estuarine osprey may be fairly common in the northern parts of its range. The white-bellied sea-eagle is much more widespread than its name suggests, being found patchily through much of the Murray–Darling system as well as far inland in the tropical north. The swamp harrier lives much of its life in and around sedge wetlands and other open, grassy places near water where it can see to hunt, sometimes nesting in smaller trees near the water.

Australian pelicans

A group of pelicans casually surrounds a black duck and her brood.

Pelicans are the most abundant of the larger, carnivorous wetland birds, and are found in large flocks in diverse habitats from marine bays to inland lakes, breeding on islands whenever enough food is available locally for them to raise their young to maturity. Clownish but charismatic, they aren't too proud to accept handouts but are also excellent at fishing communally. It took me months to photograph complete group fishing sequences at close range, and when I did it was by accident, while waiting to get a shot of a golden-headed cisticola in its spring plumage.

I eventually turned to find a group of a dozen pelicans forming into a convoy around 30 metres away and, sitting in mild sunshine with a long lens already in place, recorded a series of fishing sequences from every angle. The basic pelican fishing formation is a loose horseshoe arrangement, with the group taking its cues from the leading birds on either side. In these shallow, turbid waters it was unlikely that the hunters could see fish below, although their prey would certainly have been able to see the bulky shadows above.

As the group moves forward the open end of the formation closes together, heads tilted slightly inwards until the leaders dip their heads into the water. At the same time the wings rise and are abruptly spread, which must be a threatening sight from below; this drives fishes towards the apparent safety of the open water in the centre of the group. The prey on this occasion must all have been small as there were virtually no large fishes left in the lake, which was still recovering after years of drought (see chapter 9).

Yet the fishing must have been reasonably successful because it continued for an hour, and a group of around 30 pelicans remained around the mouth of the creek from that time onwards. I have also seen small groups of pelicans use a similar technique to casually surround ducks with ducklings, but these hunts aren't successful often and I suspect the pelicans may just be amusing themselves, for when the mother duck begins to call her warnings even very young ducklings can accelerate out of the circle at an impressive speed.

Inland shorebirds and seabirds

It is no surprise to find silver gulls around inland waters as these versatile, aggressive omnivores seem to thrive on any old muck, and have little fear of humans so they will appear anywhere with unlimited opportunities for scavenging. However, several species of terns are also frequent visitors inland, although their tastes are more ethereal and they prefer to feed on the wing over open waters in search of small fishes and larger plankton. Whiskered terns are probably most often seen, but even the much larger Caspian tern can be found in large flocks around recently flooded areas at times.

Many other birds that are more commonly associated with the sea will also move inland when conditions are to their liking, with some becoming seasonal residents, while others have evolved into distinctive inland species. These include some of the shorebirds that migrate between the Northern and Southern hemispheres, for example Latham's snipe which breeds in Asia but spends half its year on the fringes of inland wetlands in Australia. A cryptic bird that usually lurks among low grasses and sedges, it is usually seen only when disturbed, bursting into a fast, low and erratic flight before dropping to the ground again.

Various stints and sandpipers also commute globally, feeding in mixed flocks on shallow, open shorelines of inland wetlands and lakes. Sharp-tailed sandpipers are probably the most abundant inland visitors although other sandpiper species and also curlews may travel with them, making identification difficult. Other camp followers include dotterels and plovers, which find safety in numbers among the greater hordes, but are more solitary during their breeding season when they

Whiskered terns at rest among teal and swamphens.

make their very basic nests. These are rarely more than an open scrape in the ground, sometimes decorated with a few feathers or shells to help hide the heavily camouflaged eggs.

Still other types of shorebirds have evolved into permanent residents of inland waters, including stilts and red-necked avocets. These long-legged waders feed on smaller invertebrates in mud, or may pick brine shrimp from the surface of the water, but also show an unexpected aptitude for swimming in deeper waters when the mood takes them. Highly attuned to the climatic vagaries of inland Australia, they follow the rains into arid regions where they may form colonies of thousands of breeding birds near the inland saltpans and open wetlands where they feed.

The strikingly patterned painted snipe is one of the most secretive indigenous birds, with small groups appearing unpredictably in shallow, ephemeral swamps where they hide by day among low tussocks, emerging to feed over mudflats towards dusk. Now an endangered bird because so many wetlands in less arid regions (which were probably ideal refuges during drought) have been drained or degraded, it is likely that it is only their urge to move on to wetlands which may be a thousand kilometres away that has kept the species going.

Part 2
Living with change

Wetlands are often perceived as tranquil places of great beauty, and the images most likely to be seen on calendars and posters have helped to create this ideal: luminescent waterlily flowers against the background of a darkening lagoon in Kakadu; the silver sand of a shallow stream running through a Daintree rainforest; a Tasmanian mountain tarn reflecting morning sun in midsummer. Photographers avoid these places when dust from the dried-out lagoon bed gets into expensive camera gear and air temperatures rarely fall below 35°C at night, when the clear stream runs metres deep and brings down trees in its wake after hours of torrential rain, or when a south-westerly storm boiling up from Antarctica could make recovery of their frozen bodies impossible for many days.

This understandable photographic bias gives a misleading impression that comparably dramatic events such as Lake Eyre in flood, endless kilometres of heavily flowering paperbarks fringing drying backwaters from Tasmania to the tropics, and immense colonies of waterbirds nesting on islands of vegetation rising from floodwaters are somehow exceptional. Yet wetlands are among the more changeable places on this planet, and water has always been the most consistent and reliably powerful force for long-term change.

Water erodes mountains, carrying an abrasive sludge of grit and even rolling boulders along after heavy rains, dumping sediments in the sluggish lower reaches of rivers to create floodplains and nutrient-rich lowland swamps. Water has created the gentle, worn-down landscapes of Australia over more than a hundred million years, and the many species of wetland animals and plants that live in these varied habitats have had plenty of time to adapt to flood and drought.

Many of them depend upon these changes for their best breeding years, and the three chapters that follow look at the predictable ways in which seasonal changes shape the ecology of many types of wetland. A fourth chapter looks at a very different effect which can also vary dramatically from one year to the next: the way in which wetland animals may interact to improvise unexpected new ecosystems from one year to the next, all within the same physical setting.

7

Making the most of change

The smaller and shallower a wetland, the shorter its life span once dry weather sets in. Though this would seem to be the opposite of what wetland animals would prefer, some of the smallest aquatic organisms have been making good use of such changeable conditions for hundreds of millions of years. Crustaceans are often the dominant animals in ephemeral waters, partly because they have had plenty of time to fine-tune their entire life cycles to these specialised niches, and partly because their major predators are killed off for a time by the seasonal dry period, or must move on elsewhere.

Predatory insects will also arrive if the wetland remains filled for long enough, along with others such as waterboatmen that feed on organic matter and decaying plants drowned by the rising waters. And if the waters spread widely enough waterbirds will also arrive, sometimes in huge numbers, to take advantage of the bounty while it lasts. The overall succession of species in seasonal wetlands is much the same whether they are lakes or shallow rainwater pools, but the longer they last, the greater the diversity of life that will eventually appear.

Small worlds

Small, short-lived pools often swarm with tiny and mysterious animals, doing the same unexplained, wild things they have been doing for hundreds of millions of years, even when surrounded by the dullest of suburban environments. The ecology of such places is generally simple, with only a few species present, though they may also attract occasional waterbirds such as black ducks or herons that feed on worms and beetle larvae driven to the surface by the flooded soils below, in

A white-faced heron and black ducks feed around the fringes of a recently flooded pool.

search of oxygen. As soon as the worms have disappeared the avian visitors usually move on, leaving a legacy in the form of nitrogen-rich droppings that will add more fuel for the short-lived explosion of life that will follow.

A typical sequence in a freshly flooded ephemeral wetland starts with the decay of any organic matter that has built up during the dry spell. Grasses and other fast growing terrestrial plants that have established themselves from seed during the dry period will drown within a few weeks as the water level rises, breaking down to provide food for blooms of bacteria and other decay-feeding organisms. These may become so dense that they form a smoky haze in the water, and it is only under strong magnification that it can be seen that this is made up of shoals of tiny living things.

Many of the most abundant and interesting invertebrates are much smaller than the water fleas and copepods introduced in chapter 1, which can at least be recognised with the naked eye from the differing ways they swim. Wheel animalcules (rotifers) are among the most abundant of the tiny animals, and are a fine example of how far removed most miniscule life forms are from our conventional ideas of what an animal should look like.

Wheel animalcules are usually said to have a head with a mouth, a body (usually elongated), and often something described as a foot, though the planktonic species don't actually need one. Part of their fascination is that each

A shallow seasonal pool in a farm paddock, habitat for more than a dozen species of tiny animals, including rotifers.

individual rotifer is around the same size as many protozoans, the entire body of these being just a single cell, yet similarly sized rotifers are made up of hundreds of cells that form fairly complex organs. The mouth may be surrounded by various specialised feeding organs with a ring of cilia – hair-like structures that beat in sequence to propel the animal along or draw food into the mouth, creating the swirling action that gives them their common name.

Most wheel animalcules are omnivores that feed on any debris that will fit into their mouths, though some are predators in their small way, and others suck fluids from algal cells. And most of them are also females; the only real job of the much simpler males is to fertilise the eggs, though even this isn't essential in many species. Females can hatch live young genetically identical to their parent when living conditions are good, but in adverse conditions they begin to produce a different kind of drought-resistant egg which will need a dormant period before it can hatch.

This seems to be a very successful formula for survival, so rotifers are abundant in almost all types of freshwater wetlands, from the most ephemeral puddles to permanent lakes. They may swarm in their thousands in a half-empty bucket under your sundeck, serving as food for other specialists in ephemeral waters such as mosquito larvae. Even in waters that evaporate so quickly that

Springtails.

rotifers may not have enough time to grow and reproduce, there are other smaller animals, including some seed shrimps, that may appear literally overnight after rain, as all they have to do is unseal their valves and swim away. Insect-like springtails also appear within hours of flooding, drifting in grey, velvety-looking mats on the surface and feeding upon the tiny algae known as diatoms.

Larger and deeper pools last for much longer, especially in a wet year, giving time for entire communities of miniscule plants and animals to develop. Even a pool in an undrained corner of a paddock can support whole successions of animals, starting with seed shrimps which may be abundant within weeks of the pool refilling. Various types of water flea and copepod take longer to develop from dormant eggs, but some surprisingly large crustaceans can also take advantage of these relatively predator-free environments.

Crabs and crayfishes

With their hard shells and burrowing habits, it isn't surprising that many crayfishes can survive periods of drought, and some will even crawl overland during rainy periods if water quality deteriorates too much for their liking. Nearly all Australian crabs are found only in marine waters, though some have become permanent land dwellers and return to the sea only to release their eggs. While there are many types of freshwater crabs across the warmer parts of Asia, Africa and the tropical Americas, it is not surprising that there are apparently only three inland species on this generally arid continent.

Yet the single species known variously as freshwater crab, brownback or desert crab is common across most of the northern half of Australia, from perennially flowing tropical rivers to inland watercourses that may dry out for years at a time. Fishes survive in these extreme environments only in deeper, more permanent pools, but even there it may be a matter of luck as to whether they can hold out until the next rains; by contrast the crabs can seal themselves into deep burrows through times of drought. And although fishes will recolonise dried-out rivers by

Freshwater crabs mating.

swimming upstream during floods, by the time they arrive adult freshwater crabs may already be present in their thousands.

Everything about the life cycle of these desert crabs is adapted to ephemeral water in deserts, from their tolerance of a wide range of salinities to their breeding habits. Pairs seem to stay together for days before the female moults, with mating initiated by the female once she is no longer so soft that she can be damaged easily, and she tips the male gently onto his back for just a few seconds while they couple. The male guards the female through the first couple of days while her new shell hardens, but after this time she becomes increasingly aggressive towards him until he is finally driven away. The larvae of marine crabs are planktonic, going through a series of complex and often surreal-looking stages before settling down as miniature crabs, but there isn't time for this in a fast-drying desert stream so the eggs develop into miniature crabs under their mother's tail.

Many native crayfishes also show some degree of adaptation to drought, but marsh crayfishes mark a transition between the primarily aquatic species and those that may live well away from permanent surface waters. These are also known regionally as bush or black yabbies for their often discoloured shells, and with their slightly reduced tails are often confused with land yabbies (see chapter 11), which have carried the tail reduction to still greater extremes. Marsh yabbies

Slender marsh crayfish.

Slender marsh crayfish habitat: a shallow pool at the edge of the Southern Ocean.

live in shallow, ephemeral waters where they may have only a few months to forage, before having to retreat underground – sometimes for years in the event of a prolonged drought.

Slender marsh crayfish are reasonably common in diverse shallow, seasonal wetlands in southern Victoria, including in an intriguingly simplified wetland ecosystem between the ocean and the Great Ocean Road. Around 70 metres across and not much more than 40 centimetres deep when full, this pool is often a shallow dustbowl with only a few annual weeds taking root during winter rains. In a wet year it fills and comes to life, with the dominant life forms being the crayfish and the equally drought-tolerant pygmy water-ribbon. As with narrowleaf water-ribbon in south-western Australia, this extremely drought-adapted plant can grow, flower and set seed within just several months, and its tuberous roots are able to survive drought for many years.

It seems likely that the crayfish feeds mainly on the water-ribbon during relatively wet years, though it presumably also feeds on any weeds that appear during the dry months and drown later, as well as on seed shrimps and water fleas that may proliferate at times of high water. Although this may seem a harsh environment from the human perspective, there are no predators apart from the occasional visiting gulls or herons, neither of which are much of a threat for an animal that comes out only by night.

Plants: paperbarks and water-ribbons

Paperbarks are widespread through wetlands around Australia, mostly around coastal regions though some species are also found far inland. The most aquatic paperbarks (four species along northern coasts as far south as the Tropic of Cancer and also to the north of Australia, broad-leaved paperbark along much of the eastern coast, several in south-eastern Australia, and two in the south-west) may grow standing in water for months or even sometimes years, but will gradually die out if they don't have dry periods in between floodings, and their seedlings don't thrive under water. Some are associated with the fresher reaches of mangrove swamps or saltmarshes, and provide significant breeding habitat for birds and even crocodiles.

The strap-leaved, tuberous-rooted water-ribbons are often found in similar habitats but are tolerant of greater extremes including complete drying out for years, or more permanent flooding as long as the water is not too deep. Dormant plants may regrow within days of rain soaking the ground, stockpiling nutrients against later droughts in their edible tubers, but they won't flower and set seed unless they are shallowly flooded. Water-ribbon leaves and tubers are eaten by many herbivorous waterbirds, including swans, ducks, geese and swamphens, some of which also mound the leaves to create buoyant nests, and the teeming invertebrate life which shelters among the new growth of a recently re-flooded water-ribbon swamp attracts many smaller predatory birds as well as the insects they feed upon.

A seasonal swamp in south-western Australia with cutting sedge and narrowleaf water-ribbon, surrounded by swamp paperbark.

Australasian grebe foraging among common water-ribbon.

Frogs

Although usually thought of as mostly aquatic animals, many smaller frogs live in damp places for much of the year, and breed in ephemeral waters which may stand only for a few months. While the advantages of a short-lived habitat where the only likely predators are occasional passing herons are obvious, it does mean that the frogs must be ready to spawn within days of heavy rain falling, and their tadpoles must grow and turn into adults within just a few weeks. It is not surprising that such frogs and froglets are also very quick to discover new potential breeding habitats.

The first autumn downpour we experienced after moving to the Otway Ranges brought out literally thousands of eastern little froglets, and even though we live near the top of a hill around 50 pairs of these frogs arrived to mate in the shallow pool that formed around our car that very night. Apart from being an unlikely breeding pond as it was surrounded by many metres of machinery-churned mud, that pool had not existed in the previous winter breeding season, suggesting that the romantic urge in these froglets also brings out the gypsy in them.

There are dozens of froglet species that take advantage of short-lived pools, and they are found around much of Australia. It isn't hard to find their breeding habitats in eucalypt forests, grassy wetlands and sometimes even on the fringes of sports grounds by tracking their distinctive calls after the first heavy rains, and these are also a useful guide to identification as many species are so small and variably patterned that they are otherwise difficult to tell apart.

Quacking froglet habitat with south-western laceplant, a highly specialised plant of pools that may dry out within months of filling.

Location also helps with identification; for example the quacking froglet in south-western Australia sounds completely different from any other species, anywhere. In a Perth suburb I found thousands of tadpoles of this species crowded into a small pond within weeks of the first heavy rain in several years, thriving among newly sprung growths of comparably drought-tolerant plants such as south-western laceplant and narrowleaf water-ribbon that must also have sprung up from their tuberous roots in that short time, and were already starting to flower. The breeding season was far from over and the quacking was almost deafening at times, resuming all around me within a minute or two if I just stood still for a short time, and suggesting an urgency that overrode any fear of potential predators.

Even in deserts there are a few frogs that have adapted to short seasons and extremes of climate, all of them burrowers that emerge only after substantial downpours, and able to breed within hours to make the most of these rare opportunities. These aren't just a single isolated branch of amphibians that regard having a view of Uluru as more important than a reliable water supply, and there are some surprisingly similar-looking burrowers in arid regions that have evolved completely independently from the two major families of native frogs, including both the ancient Gondwanaland group and much more recent arrivals from Asia.

Fishes

With their strong ties to water, fishes are the most sensitive of aquatic animals to drought, yet some species, including many of the pearlfishes of South America are mostly found in ephemeral pools and other unreliable waters, growing to maturity and breeding within several months to produce tough eggs that hatch when the next season's rains refill the pools they live in. Others such as the African lungfish will burrow into mud and can recover even when their bodies have become partly desiccated. There are many comparable habitats in Australia, yet you can count the

Salamanderfish in the peaty water of their preferred habitats.

indigenous fishes that can survive drought on the fingers of one hand – though you may need both hands before long as genetic studies turn up new but so-far cryptic species.

The dwarf galaxiids of south-eastern Australia (see chapter 14) can certainly aestivate for a very short period during dry seasons, though other evidence suggests that they are more likely to retreat into hidden sumps such as crayfish burrows, and from my own unplanned experiments in nursery ponds it seems unlikely that they will survive for long in soils that aren't actually boggy. The Tasmanian mudfish is known to tolerate fairly dry conditions for long periods of time, even in damp soil under tree stumps, but unlike its New Zealand relatives is usually found near coastal streams as it must return to the sea to breed.

The salamanderfish of south-western Australia is the most thoroughly drought-adapted native fish of them all, a slender species with a flexible neck that makes it seem most un-fishlike as it creeps across the bottom of ponds, turning its head in search of prey as it is unable to focus its eyes otherwise. Where it came from or what it evolved from is not known, but it is almost certainly an ancient Gondwanaland fish that has had perhaps 90 million years to adapt to increasing aridity as the continent drifted northwards.

Originally thought to be related to the similarly shaped but scale-less galaxiids, the salamanderfish became one of our most-studied inland fishes as its many

Salamanderfish habitat: a borrow pit in acid heathlands.

unique and curious abilities became known. These include the ability to burrow when the dark, peaty and acidic ephemeral ponds that are its usual habitat dry out. Water loss during this time is reduced by a slimy sheath that forms on the body, and its metabolism also changes while it aestivates until the rains return. Although restricted to a relatively narrow coastal belt this species is common in many places, including well-protected areas in national parks, but it is also a very adaptable beast that readily colonises new places such as borrow pits where gravel has been extracted to make roads, as long as these are in sandy soils with suitably acidic water.

Birds

With the advantage of flight, birds can move on as required in search of food or breeding habitat, so the ephemeral nature of many wetlands poses few problems for most of them. Some species actively seek out recently flooded wetlands where tadpoles, insects and larger crustaceans may be seasonally abundant, while ibis are attracted to temporarily waterlogged paddocks or lawns, where worms and beetle larvae must surface to breathe; in deeper floodwaters they will be replaced by herons and ducks. Others may follow drying watercourses in search of shallow pools where fishes are trapped, but most of these species are just opportunists,

Brolgas.

taking advantage of a temporary bounty while it is exposed, rather than specialists that rely on change for their living.

Brolgas are the largest of the many types of native waterbirds whose entire life cycle is based around seasonal change. These tall and elegant birds with their exotic dances and pensive expressions are primarily vegetarian, although they also feed on frogs, large insects and even some molluscs. They seem to lack any fish-hunting skills, presumably because fishes are rare in the shallowly flooded open places where they breed. Brolga nests are usually built on low rises or are raised above the water, made from plant materials held together by mud, and are literally surrounded by a freshwater sea of high-protein foods right at hand, allowing the chick (or sometimes even two) to grow to flying age within as little as 3 months.

Even outside the breeding season when their protein needs are much lower, brolgas in relatively undisturbed habitats are still dependent on seasonal change. In the north they feed largely on bulgurru, a native sedge that forms abundant crops of small, slightly sweet, starchy tubers as the shallows where it grows dry out; Chinese water chestnuts are a larger selection of this plant. In southern Australia where brolgas have become much less common with drainage and general mismanagement of ephemeral wetlands, other tuber-forming sedges they once relied upon have become much less abundant since the advent of grazing, and they have taken to grain as a staple of their diet instead.

Heads under, a pair of pink-eared ducks swirl as they bring plankton up from the shallow floor of a lake, with black-winged stilts picking at the larger animals that got away.

Some waterbirds are more specialised feeders on planktonic animals, and their breeding success depends on recently flooded areas of water that are not too deep, with suitable nesting areas not too far away. Zebra ducks (better known as pink-eared ducks, though their so-called ear is tiny and almost impossible to see at any distance) are among the most specialised of these, and though they are not uncommon on larger lakes during periods of drought these are probably just refuges for them, places to wait out hard times until the rains trigger their restless migrations. Their striking patterning and strangely shaped bills also come with some novel behaviours, including one which has been described as courtship, with pairs of the ducks whizzing around in circles with their heads under water.

Given that the water they are in is often murky, this would seem a strange way to show off their dramatic plumage to potential mates, but it makes a lot more sense if you have ever collected plankton with a net. Water fleas and copepods aren't spread uniformly through water; they move in pursuit of phytoplankton, and phytoplankton rises and falls over the course of a day, presumably adjusting levels to make the best use of light and temperature. If there is a breeze, zooplankton move down towards the bottom where turbulence is minimal, or concentrate along the lee shore of higher ground where the surface is less choppy.

If you use a fine-meshed dip-net you will quickly find that the best way to bring the concentrated zooplankton layer up from the bottom is to swirl it in a figure-eight pattern, creating a vertical uplift that carries the tiny animals to the surface. Zebra ducks can't do a figure-eight but create the same effect by swimming fast in pairs in a tight circle, while their beaks work through the column of rising water. Although this method can also help to bring up zooplankton from deeper waters in a large lake, it is far more efficient in shallow waters where plankton lies in a particularly concentrated layer just above the silty bottom.

When plankton densities are extremely high, as in recently flooded shallow wetlands of any kind, a vigorously wielded plankton net can be choked within seconds, and the ducks can probably pick and choose whatever they want from the rising waters. A sure sign that the pickings are good is when other waterbirds such as stilts are seen working their way around a pair of swirling zebra ducks, selecting from the insect larvae brought up with the smaller animals.

The plankton that may swarm in immense numbers in recently flooded wetlands attract smaller predators as well, of which backswimmers are usually the most abundant in waters where fishes have been killed off by recent drought. Waterboatmen are also likely to be common along with bloodworms, and all of these make nutritious foods for other specialist birds such as hoary-headed grebes, which may also breed in prodigious numbers in such places – but that is a story to be told in chapter 9.

Human impacts: drainage

We are living in a dustbowl now compared to what this country was like when the first European settlers arrived, in an increasingly arid wasteland that is largely of our own creation. In many areas up to 90% of wetlands have been drained, and even the terrestrial vegetation around the disappearing wetlands has been changed by selective grazing, compaction by the hooves of livestock, and weeds introduced with so-called improved pastures. Once covered in native grasses, pigfaces, yam daisies and other edible plants that not only fed Aboriginal peoples but also created open-structured, moisture-retaining soils, many inland areas are now permanently drought-affected to varying degrees.

In the eyes of most landholders, there appears to be a strange conflict between the urge to drain areas which remain wet or shallowly flooded for part of the year, and valuing them highly as a source of lush, green fodder when the surrounding pastures have turned to hay. The first European settlers brought this unfortunate habit with them from countries where rainfall is frequent and evaporation is low, but in Australian conditions it has only accelerated the degradation of the wider environment. And although it would be hard to prove, the reduced humidity associated with drainage has probably had an effect on the reliability of rainfall as well.

If it is any consolation, the damage is worse in many other 'developed' parts of the world. Consider Florida in the USA, once one of the world's most extensive

This single urban drain has dried out around 3000 hectares of swamps and other wetlands.

wetlands, which is now so thoroughly drained and manipulated that 90% of the water used by the population spurt of the past few decades must be pumped from underground – and even this supply is running out very rapidly. One of the worst weeds in Florida is the eastern Australian broad-leaved paperbark, a significant habitat plant here, the seed of which was scattered so widely across Florida in the hope that it would help to dry out wetlands that it will never be possible to eradicate it.

8

Moving on

The simplest way of dealing with dramatic changes in wetlands is to simply move on – if you can. Animals that can fly have no problems escaping a drying water body, while many frogs, turtles and even some crayfishes can walk or hop reasonable distances, and even wetland plants can move surprisingly fast to take advantage of new habitats, as will be seen in the next chapter. The drought-tolerant, dormant eggs of many smaller crustaceans are so small they may blow with the wind, or their hard shells can survive passage through the guts of a waterbird. And as the next generation of many smaller crustaceans will ultimately produce both males and females from any single female that hatches, it takes only one egg to colonise a new habitat.

In the case of some other widespread freshwater invertebrates we have little idea of how they move between one wetland and another. Consider glass shrimps; widespread and often extremely abundant throughout south-eastern Australia in diverse habitats, they are considered to be true freshwater animals, though tolerant of some degree of salinity. Yet they are also present in the isolated coastal streams of the Otway Ranges, which were probably often dry even a few thousand years ago. Have the eggs of berried females made their way to these young streams via the digestive system of fishes, or perhaps waterbirds? Or have they managed to survive in isolated pools of freshwater that were gradually pushed back by the rising sea, as Bass Strait flooded after the last ice age?

Fishes and the upstream urge

Although it is sometimes suggested that fish eggs can also be transported by waterbirds, there has never been any convincing evidence that this has ever

A 'berried' (egg-carrying) female glass shrimp forages for algae on river pebbles.

happened with native freshwater fishes, which can only conveniently move between diverse types of wetlands through the rivers, streams and backwaters that may link them during wetter times of the year. That doesn't necessarily mean that they follow these routes voluntarily, as some fishes are specialised enough that they are likely to colonise new places only when swept from their original homes by floods.

These were the aquatic highways through which Australia and New Guinea shared many fish species during the last Ice Age, when Torres Strait was a plain that joined the two great islands into the still larger landmass known as Sahul. However, many others willingly move long distances upstream into unknown waters where even the deepest pools may dry out after a few months without rain. This impulse to swim upstream against floodwaters makes little sense to casual observers, as the great majority of fishes that make their way into these unpredictable waters will die without breeding; on the positive side they create easy fishing for herons, pelicans and cormorants as creeks and waterholes run dry.

Yet this is a successful strategy for many species of freshwater fishes when seen from an evolutionary perspective, and the urge to move against the flood is so strong that substantial schools of catfishes, perch and others may even be found swimming on their sides if floodwaters become too shallow for normal progress. I use the term 'fresh water' loosely here, and this is as good a time as any to remind readers that in the arid inland many of these fishes must be tolerant of an impressive range of conditions from sweet and fresh to bitter waters saltier than the sea.

Spangled perch are the best example of a fish that compulsively migrates against the flow, and with a range covering most of the northern half of the continent this is the most naturally widespread freshwater fish in Australia. They

Spangled perch.

have sometimes been found stranded yet still alive within hours of a heavy downpour, many kilometres from the nearest permanent waters, helping to perpetuate the myth of fishes that come down in rainstorms. Stories of frogs that have apparently come down with rainfall are less surprising, as there are also many burrowing frogs that may appear in large numbers after a rain.

Spangled perch aren't just some kind of dumb automaton that barely manages to survive; they are among the most intelligent and adaptable of inland fishes. A perch as long as my hand will happily spend hours snacking on individual water fleas, something most other comparably sized fishes can't do because their eyes are designed to focus only on the larger animals which are their primary prey. And despite a reputation for murderous aggression towards other fishes as well as their own kind, young spangled perch will also school to herd potential prey into a small volume of water, where each fish can feed more efficiently than if their victims were more randomly spaced.

The apparently suicidal behaviour of fishes that swim against the occasional flood makes more sense if it is considered as a type of lottery, with just a few big-time winners. Even if the great majority of each generation that swarms upstream is eaten or just ends up suffocating in the thickening mud of a drying pool, their fate has no effect on the genes of the survivors. The future breeders are the few lucky individuals that find a new home from which they can spread their own genes, and *just because* they are the lucky survivors their offspring will inherit this urge.

The upstream urge isn't confined to one or two similar types of fishes, but is a potent force in several unrelated groups. These include two types of eel-tailed catfishes, the golden perch (though the desert forms may be a separate species to the Murray–Darling form), several grunters and glassfish. The freshwater herring known as bony bream can be so abundant in inland waters that it has been commercially harvested, and was tinned as a staple food for troops during World War II. True to their name these fish were so bony that my father-in-law still contemptuously remembered this gourmet treat in his combat rations decades later as 'goldfish'.

The salt tolerance of most inland fishes is no great surprise given that the ancestors of every single species found in these harsh conditions has evolved from marine ancestors, but desert waters fluctuate wildly from nearly fresh at times to seawater strength, and their temperatures can rise from near-freezing to an oxygen-starved 40°C over the course of a single day. As nearly all native fishes found in desert areas originated in the warmer waters of the north, the process of adaptation must have taken a long time while they developed the broad tolerances needed for some of the most unpredictable habitats for freshwater fishes on Earth.

We can see the early development of this upstream urge in many of the fishes of the Top End, where the extensive flooding caused by monsoon rains encourages dispersal across the great floodplains, including many pools and billabongs that will dry out completely as the dry season progresses. Rainbowfishes are among the most abundant and diverse freshwater fishes of tropical Australia, partly because

An ephemeral creek in the Top End, habitat for several species of rainbowfishes.

Western rainbowfish.

they must breed prolifically to keep their numbers up, and partly because they are a popular snack with every predator with a taste for fish, but also because they spread so widely during the Wet that literally billions of them must be doomed every year when receding floodwaters leave them trapped.

This strategy was easily extended into more inland waters, with the western rainbowfish following river gorges deep into the arid zones of the north-west, while the desert rainbowfish can be found right into the heart of Australia – in a good year! The rainbowfishes of south-eastern Queensland have travelled even further as they followed the Darling River southwards, then turned upstream through the still colder waters of the Murray River and its southern tributaries.

The change in water temperatures and other conditions where the Darling joins the colder, mountain-run-off waters of the Murray must always have been abrupt, so upstream progress must have been slow for this formerly tropical species. It may have died out in the cooler reaches of the Murray time and again during centuries when water temperatures were cooler than average, but the survivors eventually reached as far south as central Victoria, evolving along the way into a blunt-nosed, golden-green form that looks very different from its crimson-spotted relative around the headwaters of the Darling River.

It isn't just fishes that have the upstream urge; some crayfishes are also triggered into restless motion by increased flows, particularly redclaw from northern Australia. This large species has such a powerful migratory urge that it will climb a ramp with water running down it, a habit used to advantage in traps where the climbing animals fall into a harvesting box as they leave their dam. The downside in environmental terms is that redclaw escapees from aquaculture farms are making their way inland along ephemeral rivers that were relatively pristine until recently, with unknown and unpredictable consequences for their future ecology.

Origins: moving inland from the sea

Nearly all freshwater native fishes are descended from marine and estuarine groups that have moved inland, so it is not surprising that many can tolerate impressive levels of salt. And quite a few freshwater fishes still breed in estuaries, or their young may move out to sea for some time, which is why all but the smallest and most ephemeral coastal streams are likely to have their own fish fauna. Given the generally arid and unpredictable nature of this continent, it is likely that fishes which can move between fresh waters and the sea will continue to dominate many inland waters until Australia has run aground on Asia, and new families of true freshwater fishes have their chance to move south.

Tupong.

The tupong (also known as congolli or freshwater flathead) is one of the most intriguing pioneers from the sea. A freshwater cousin to a mainly Antarctic and sub-Antarctic family (including icefishes that aren't bothered by temperatures below freezing point), it is so tolerant of change that you can transfer it from seawater to fresh with no apparent effect on its health or appetite, though it usually stays within striking distance of the sea where it is believed to breed. We still have no definite knowledge of the life cycle of this or of many other marine and freshwater species, however abundant they may be.

Other southern families of fishes such as the galaxiids are more flexible, with some species able to breed in fresh water if trapped inland (see the next chapter). In the warmer waters of the north many other fish families include species that regularly venture inland. Some such as the diamondfish are primarily marine, though young fish may move into freshwater streams for weeks at a time – a case of testing the water for future generations, perhaps? Various sharks, rays, sawfishes, toadfishes, bream, several pipefishes and even a moray eel are also known to move into fresh waters regularly.

Diamondfish.

Overland or by air

Some frogs and reptiles are excellent migrators; the tracks of snake-necked species such as the oblong turtle in the south-west and the common snakeneck in the east are common on drying expanses of mud in warmer weather after rain, and both may feature prominently in roadkill statistics as they tend to tuck away into their shells as a car approaches rather than running for their life.

However, it is not so obvious that some frogs will also travel great distances during wet weather as they are far more secretive. The brown treefrog of south-eastern Australia is a great coloniser of new habitats, including farm dams, once

Brown treefrog hunting moths on a warm, wet night.

there is enough vegetation established around them, and can also be found in permanent ponds many kilometres from the nearest wetlands even when there are never connecting waters in between. They seem to appear in hundreds on some rainy summer days, their strident, repetitive calls jarring the night as males sing from inside our water tanks while the females make themselves known by swarming across the windows in search of moths.

In a sense, frogs could be described as a sophisticated delivery system, capable of introducing their eggs (and ultimately tadpoles) to waters where predators are less likely to be abundant. And the tadpoles need all the help they can get along these lines, because they are among the most popular and easily caught food items on many wetland menus, being mostly slow, bumbling and loaded with protein. The reason so many species of common frog produce hundreds of offspring every breeding season is because, in most places, most of them will be eaten long before they have the chance to turn into frogs.

In recent decades, many frogs, including several indigenous species, have disappeared completely from the wild, and the primary culprit is suspected to be chytrid fungus. In response, many concerned individuals have set up specific frog ponds in gardens and urban areas, but these may just be exacerbating the problem.

Brown treefrog tadpoles from a small pond several kilometres from the nearest stream.

Almost none of the vanishing species are ever found near cities; instead it is the common species that are largely immune to the fungus that colonise and breed garden ponds, and the gardens that surround these predator-free waters are almost invariably too small to support the hundreds of frogs that may emerge every year. With insufficient food at hand, most of these frogs are likely to move on, potentially acting as carriers for new strains of chytrid fungi and other introduced diseases that may be affecting wild populations.

Insects

Travel by foot is slow, and there is a limit to how far it can take an aquatic or semi-aquatic animal, particularly during drier weather; the best way to move rapidly between widely spaced wetlands is flight. It is an energy-expensive option in many ways, and not available to animals such as crustaceans that can breathe only through gills, though their fellow arthropods, the insects, get around the problem by only having gills during their aquatic larval stages.

The most common aquatic insects are invariably excellent colonisers of new waters, sometimes appearing in recently flooded wetlands in such impressive numbers that there must be thousands of them up in the air on suitable nights, even though humans rarely see them. Bugs such as backswimmers and waterboatmen will often appear in smaller garden ponds within weeks of filling, but even the

Orange-headed waterbeetles may appear in farm dams within a few months of them filling.

largest waterbeetles may sometimes appear in their hundreds on a warm night when the urge to migrate comes upon them, hammering on windows so they sound like hail in summer, and crawling clumsily underfoot as they try to take off again.

Just because an insect *can* fly doesn't mean that it is likely to travel any great distance in its lifetime; accomplished fliers such as dragonflies are a notable example, as males often just claim a restricted territory around a body of water that is likely to appeal to females of their species, and defend it against all-comers. This is partly why many species of dragonflies and damselflies are surprisingly restricted in range, staying around highly specific types of habitat and within limited geographic areas.

There are dragonflies that breed only in peaty dune lakes, others that stay around deeper pools in fast-moving streams, and some smaller species that even stick to boggy seepages. Yet many other dragonflies are found across most of Australia, including in desert areas during wet years, and I have seen massive flights around recently refilled lakes that are so dense they give the impression of a swarm of locusts at first glance.

Waterbirds – moulting and migration

Of all flying animals, birds are the greatest travellers, but as this ability is tied closely to many aspects of wetland biology discussed elsewhere in this book, it need not be considered in any detail here apart from some of the less obvious mechanical issues. Most ducks seem well designed for fast flight, their choppy wing beats lifting them rapidly until they settle into a steady pace that can carry them many kilometres without a break. With their broader wings swans need to work hard at take-off but soon fall into a slower rhythm as they choose their altitude to match the distances they intend to travel.

By contrast, the wings of more predatory aquatic waterbirds such as cormorants and grebes are not very efficient as their body design is primarily arranged for speed underwater, so grebes prefer to fly by night when harriers and other raptors are asleep. Cormorants don't mind flying by day but are sluggish at lift-off, and those that visit our swimming dam may need to spiral through half-a-dozen circuits until they have gained enough altitude to clear even the relatively low scrub and bush growing well back from the shoreline.

Pelicans must also work hard to get aloft but are superb gliders, and with their broad wings they can take advantage of thermals to do most of the lifting needed on a sunny day; once they have gathered speed they can sail great distances or skim smoothly just above the waves. Herons and egrets are the most graceful fliers to my mind, with legs stretched straight behind and necks tucked in to create a handsome

Great egret in flight.

profile. The great egret should be the most graceful of all, but spoils the effect with a conspicuous kink in its recurved neck that makes it look as though it has swallowed a box of matches.

Feathers are the defining component of bird flight, and although these are well designed for light weight and easy repairs after minor damage, they need to be replaced regularly as they age. Many birds moult over a fairly long period of time, shedding their old feathers according to a fixed pattern, so they don't lose the ability to fly even though they may be reluctant to travel any great distances during this time. It takes a lot of food and energy to grow a new set of feathers twice a year (or even just once!), so moulting is usually timed to fit in with availability of an adequate food supply.

Many ducks moult and replace their feathers so quickly that they become flightless for a time, particularly the Australian shelduck which often gathers in large groups for its post-breeding moulting season. During the pre-breeding moult they lose only part of their feathers, replacing these with the showy plumage associated with courtship, but during the next and more prolonged moult shelducks take nearly a month to regrow enough flight feathers so they can move on to new pastures, separating into smaller groups and pairs at this time.

All waterbirds must follow the rains to some degree, some moving predictably according to seasonal patterns, but in Australia many endemic species are better known for their erratic appearances over a wide range. Some ducks such as hardheads, whistling ducks and pink-ears are particularly noted for their restlessness,

A mixed flock of hardheads and teal.

and it is usually assumed that they are so well attuned to the eternal cycle of drought and flood that they just can't help themselves: the idea of 'walkabout' applied to birds. Hardheads are particularly flighty and have formed breeding colonies as far away as New Zealand at times, though the Murray–Darling seems to be their particular stronghold at the present time.

The search for new sources of food is obviously part of the motivation for their migrations, yet as will be seen in the next chapter the impressive congregations of waterbirds that appear around recently refilled wetlands may move on long before the food supply is noticeably reduced. It is not just an isolated phenomenon, as during the early stages of photography for this book I made a lengthy trip through some inland areas in pursuit of the waterbird concentrations that had been reported from many places between 3 and 4 months earlier. The great quantity of smaller animals, from planktonic species to insects and larvae, was still impressive in many of these places – but so was the virtual absence of waterbirds apart from the odd scattering of ducks, egrets and the occasional cormorant.

Perhaps it is the sheer numbers of waterbirds that build up under good conditions that are the problem in themselves, as this must be the ideal scenario for parasite build-up as well. Many of the smaller aquatic animals are not just food; they are also likely to be intermediate hosts for diverse parasites, some of which have been studied, though many others are poorly known or even still unnamed. With the many treatments now available for humans and domestic animals, it is easy to forget that as little as a century ago many parasites were largely untreatable. In the case of waterbirds the best option for parasite management is to keep moving on, before the parasites build up in such numbers as to seriously affect their hosts.

Human impacts: climate change

Many indigenous wetland animals are well adapted to periods of prolonged drought, and also flooding, while others simply move on to wetlands elsewhere if they can. However, the problems they face have increased considerably since European settlement, especially as extensive drainage has made the distances between wetlands much greater in most 'developed' areas. This isn't so much a problem for migratory waterbirds, most flying insects or even those animals and plants that can hitch a ride, but it has certainly made it much more difficult for frogs and turtles to move elsewhere except during periods of extensive flooding.

The new threat looming is the unpredictable nature of climate change, with both droughts and floods predicted to be increasingly severe and prolonged in many parts of Australia, along with more gale-force winds that may keep birds from feeding for days at a time. With refuges further apart and increasingly hemmed in by cleared, open lands and housing, the adaptations that have carried most of these

Two species of egrets hunker down in the face of gale-force winds.

organisms through similarly turbulent times in the deep past may no longer be enough to ensure their survival. However, it may be humans who will be among the earlier casualties if the most extreme forecasts come true, particularly if agriculture fails, leaving the survivors to find a more balanced role in the natural world of the future.

A shallow stream bursts its banks after torrential rains.

9

Rebirth of a lake

The first time I saw Lake Colac and the even larger Lake Corangamite was as a zoology student in the 1970s, at a time when Ian Bayly and David Williams were leading a renaissance in Australian limnology: the study of inland waters and their ecology. The south-western district of Victoria was one of their prime research areas with hundreds of lakes both small and large, ranging from fresh to hyper-saline; and with a wealth of species as varied as anything that can be seen in a comparable area elsewhere in Australia it was also a very convenient place to take their students.

A few years later my wife and I moved to the Otways, and as Colac is our nearest substantial town I have spent thousands of hours around its lake over the past three decades, observing and photographing, including through the prolonged drought that ushered in the 21st century. These changes were not just a microcosm of the effects of drought across southern Australia through that decade, but also of the still greater changes that wetlands across the whole of this country must have experienced many times, long before European settlers were present to record them.

Left stranded in a shallow depression when the lava wall of the original Lake Corangamite was partly breached, Lake Colac has come and gone over the centuries, as a lunette of wind-heaped sediments to the east testifies. In wet years gone by, the lake would overflow across marshy ground, or reverse along its main feeder stream to discharge into the Barwon River, and this flushing effect helped to keep salinity levels moderately low. Early studies of the lake's biota suggest it has changed little over the past century, with only a few larger sedges, reeds and sometimes water-ribbons around the fringes, and small-fruited water-mat forming submersed thickets during times of high water.

A heron drifts across the rising waters of Lake Colac, swarming with damselfly nymphs before the return of fishes.

It needs only a slight breeze to keep the shallow waters of Lake Colac stirred over its 25 square kilometres of surface, so the lake is usually oxygen rich except in pockets where pollutants from industry and a sewage treatment plant have sometimes concentrated, though these problems are now much better regulated. Bays and inlets along the southern shore of the lake provide shelter from different wind directions and add a little plant diversity in places, while a shallow, created wetland not far from the lake was declared a bird reserve some time ago, to become a valuable refuge and breeding ground for some waterbirds during the height of the drought.

Drought and the new millennium

Cycles of flood and drought are frequent in more arid regions of Australia, but it is only when these extend more widely and affect better-watered places closer to the coastal regions where most Australians live that they are really noticed. Arthur Upfield's novel *Death of a Lake* is set somewhere in the arid inland, and his sparse and elegant descriptions of a dying lake which will one day fill again when floods move down from the far north make this his most interesting work.

The dying glory of Upfield's unnamed lake lacked something even more spectacular: the great waves of living organisms that surge into a wetland brought back to life by rain or flood. For the first decade of the 21st century, south-western

Carp skeleton marking a biological boundary.

Victoria (along with many other parts of Australia) was in a state of more or less perpetual drought, and as Lake Colac's waters fell through this time, the once moderate levels of dissolved salts became increasingly concentrated. Eventually the increasing salinity reached lethal levels, killing off most of the carp which were the dominant fish at that time; their bodies washed ashore in thousands to form a ring of skeletons that marked a biological line beyond which this introduced species cannot pass.

As the water level dropped year by year, newly exposed areas of lake floor were covered by neat bands of plants starting with seedlings of marine clubrush and sharp clubrush, germinating in huge numbers not far from the old shoreline, along with golden-headed clumps of variable groundsel from seed spread around by the wind. Other plants, including blown grass, willow herbs, water buttons and the introduced bushy starwort, became dominant further out as the waters fell further, all arriving as wind-blown or floating seed that sprouted freely on the newly exposed, nutrient-rich soils and grew fast.

By January in 2009 the lake had almost disappeared, leaving only a small, stagnant lagoon just a few centimetres deep, replenished (if that is the right term) by a trickle from Barongarook Creek where dead ducks floated not far from the creek mouth. The bird reserve was still shallowly flooded, fed with fresh waters from urban run-off, and in the autumn of 2009 around 120 black swans congregated there to compete for nesting space, as nearly all other wetlands in the area were dry.

Black swans crowd the bird reserve.

The chaos was unbelievable at times as around 25 pairs of swans attempted to construct nests in an area of just 3 hectares, but ultimately only the six most aggressive pairs succeeded in laying eggs and raising young. Flocks of cattle egrets, black-winged stilts, purple swamphens and diverse ducks also roosted on the many abandoned swan nests through this peak period of the drought, creating a living tapestry that seemed to change almost every day. The only constant through most of this period was the endless calling of clamorous reed-warblers in the reed beds along the northern fringe, experimenting with endless repetitive variations on their simple, fluid songs.

Plants: movement

Many aquatic plants spread by runners follow rising and falling water levels, and others take root from fragments broken off during floods, but a surprising number of plants follow changing water levels by growing from seed. This was particularly obvious in Lake Colac through the years of drought starting around the year 2000, as huge numbers of seedlings of several of the most common sedges appeared on the newly exposed mud. Every year their parent plants along the original shoreline had been producing millions of viable seeds, most of these sinking into the dark waters further from shore, where they were unable to germinate until exposed to light and air.

Sedge seedlings came up like a lawn in many of the more sheltered bays, growing so fast on the nutrient-rich soils that they were producing seed crops of their own within not much more than a year. As the lake dried out further the sedges were joined by new swathes of plants, most of them arriving in the form of wind-blown seed. Blown grass, willow herbs and variable groundsel formed their own mosaics over parts of the lake floor that were too far from shore for the heavier sedge seed to reach, flowering and seeding heavily for a short time before disappearing under the returning waters as the drought relaxed its grip.

As the lake continued to refill, many of this most recent generation of sedges drowned, but their seed still lies dormant in the lake bed and may last for a decade or more. Some may be disturbed by storms that wash the seed ashore where it may

Not long before the end of the drought, diverse species of plants formed neat bands and circles on the nearly dry lake floor, a living map of changes over several years as the lake dried out.

either dry out or grow, according to the climate of the time. For each species as a whole it doesn't matter which seeds succeed in any generation as long as there is a wide range of variability in their responses to drought and flood, so that no matter what this continent throws at them, the next generation will be able to respond in many different ways.

Returning waters

Moderate rains in autumn 2009 brought a new flush of growth to the parched lake floor, and within weeks most swans in the bird reserve had fled the self-righteous rage of the established breeding pairs, spreading out across these succulent green pastures instead. Though the lake itself was still just a large but very shallow puddle, minute planktonic algae and the zooplankton that feed upon them must already have been proliferating. In waters too shallow even for a kayak and too deep in gelatinous mud for safe wading, we may never know exactly what was going on in that distant pool as it moved from place to place over the lake floor, shifting almost every day according to the whims of the wind.

The two main tributaries of the lake (Barongarook and Deans creeks) were also refreshed, and though it would not be obvious until summer late that year some native fishes were making a dramatic recovery in their running waters. With a long

Adult spotted galaxias.

history of dry periods the lake has never been much of a haven for native fishes, though several species, including flatheaded gudgeons, would thrive in variable numbers. Surprisingly, it was two fishes that normally breed in estuaries and move out to sea for the first few months of their lives that made the most dramatic recovery, and by the end of that year the turbid stream was alive with large, uniformly sized schools of galaxiids.

As their spawning is triggered by rising waters neither could have bred earlier than autumn of that year, yet the young common galaxias were already around 5 centimetres long by that time, while the spotted galaxias ranged from 7 to 8 centimetres in length. These uniform size-classes suggest a single spawning event that must have happened as soon as there was enough running water in the creek, and the impressive lengths they reached in such a short time must have been because of the abundant food supplies already in the lake. The barred juvenile markings of the spotted galaxias seemed to confirm this, as coastal fishes of a similar size caught just 50 kilometres away already had the fully developed spotted pattern typical of adult fishes, so these were probably older and more mature despite their relatively small size.

Is it possible that many thousands of galaxiids could have grown so large in around 7 months? A few decades ago there were still debates about whether the marine stage of common galaxias (see chapter 17) could grow to the size commercially harvested within just 6 months of hatching, as they returned to fresh waters, but the doubters proved to be spectacularly wrong as the sea is a rich feeding ground for a fast-growing fish. So land-locked populations with an abundant food supply must also be able to take advantage of such largesse, and the food needed was certainly there as the next 2 years turned the recovering lake into a soup of biological activity.

Within a few weeks of galaxiids appearing in the creek pelicans also returned, with a group of around 30 basing themselves around the creek delta. Local lore has

Worms, bloodworms, copepods and other minute animals from the lake.

it that there were two populations of pelicans: one group that lived and fished as pelicans always have, the other living mostly on what they could find at the tip. Pelicans are opportunists that feed on anything large enough to seize and swallow, but contemporary tips are much cleaner than the dumping grounds of the past. It did not take long to confirm that the birds that had taken up residence were making most of their living from fishing: with a little help from the local commercial eel fisherman who would throw them a bucket of small carp from his fyke nets every day or two.

In late winter 2010 rain returned with a vengeance and the lake rose rapidly in just a few days, though it was still only half as deep as it has been in the past. With the returning fish populations still greatly diluted in that considerable volume of water, and floodwaters disturbing nutrient-rich sediments, a new surge of life began. Among the drowning vegetation well out from the shoreline hoary-headed grebes must already have been nesting unseen in large colonies, while below the surface millions of damselfly larvae were growing fast, feeding on the myriad zooplankton and other small animals that jerked, hopped or slid past their mask-like faces.

All of this activity was fed and fuelled by the phytoplankton proliferating in the top few centimetres of water, because further down there wasn't enough light for any kind of green plant to grow. Copepods soon became by far the most

Mussel-ostracod, a type of seed shrimp.

abundant planktonic animals in the lake, but a pass with a fine net would also bring up a few water fleas, and even the occasional relatively large mussel-ostracod speeding on its restless, erratic path. Smaller seed shrimps mumbled their way through the soft sediments on the lake floor and scuds were also common, racing from the shelter of one empty pea mussel shell to another.

The living pea mussels were deeper in the soft sediments, along with rust-red worms of several species, and bloodworms in vast and uncountable numbers. Within a few months these would emerge as chironomid flies, a non-biting midge so abundant that they would literally cover every surface within a short distance of the lake. For the few short weeks of their lives it became impossible to take a photo that wasn't just a little fuzzy from the living mist they created, or to breathe without inhaling a few. Males are impressive with their intricate comb-like antennae, clinging to every suitable surface in their trillions, waiting for a potential mate to pass nearby.

Other aquatic insects were also common, particularly backswimmers which fed directly on the copepod bonanza, becoming so abundant that windswept heaps of their shed exoskeletons formed mounds along the shoreline at times. Waterboatmen were even more abundant over the softer sediments, and the tangles of blown grass that washed ashore were often heavily weighted with their stalked eggs. The

Male chironomid flies, a non-biting midge.

adults must return to the surface to breathe at times, but newly hatched waterboatmen are small enough that they can get nearly all of their oxygen needs directly from the wind-churned waters. These tiny babies seem quite content to rummage through the bottom silt for hours, their only movements an occasional swirl of their 'oars' as they settle face-first into the muck.

Over the course of the summer starting in 2010, the dense band of living plants around the lake's edge came alive with damselflies and dragonflies, leaving thousands of their empty husks clinging to the sedges still standing above the shallow waters. In turn, the swarms of blue, red and green damselflies brought their own predators with them, from praying mantises almost as long as my hand to the long-jawed spiders whose webs seemed to be strung between every clump of rushes at the water's edge, glittering with damselfly wings.

As the weather cooled the damselflies disappeared and there would be almost none along the lake shores in the following year, with the waters below swarming with the introduced plague minnow by that time. But the stage had been set for a new explosion of life in the spring to come, as everything that any waterbird could wish to feed upon was now present in huge numbers. The water was clean though still very turbid, fishes both native and introduced could be caught with random dips of a net even well out from shore, lush pastures of edible herbs and grasses

Common blue-tails mating.

fringed the lake, and even the most ephemeral streams around the lake were flowing steadily.

The feathered hordes

Towards the end of spring in 2011 waterbirds returned to the lake on a scale that had not been seen for decades, with new species arriving in flocks of thousands all through the following summer, sometimes just days after the previous arrivals were starting to become a familiar part of the landscape. The coots that had first arrived in 2008, feeding across the shallowly flooded herbfields of the drying lake, were still present in fluctuating numbers, along with small flocks of black-winged stilts that broke up into scattered pairs as the waters rose higher.

Other birds, including sacred ibis and both species of spoonbills, had also come and gone on a smaller scale, although a few odds and ends of these species could always be found if you were prepared to walk far enough along the shoreline. The purple swamphens that were often abundant in the bird reserve from mid-winter 2008 had also come and gone in flocks of up to a few dozen, apparently just wandering between different wetlands in the area, and sampling new places or plants as the rising waters flooded them.

Cormorants, herons and egrets reached their greatest numbers over the course of 2010, and combined with the return of a fairly substantial flock of pelicans earlier that year this was an indication of reasonable fish populations in the lake, though the cormorants generally preferred to fish in the somewhat clearer waters of the creeks. The cattle egrets that had roosted in and around the bird reserve through the worst of the drought years were now spending their days further afield among herds of Friesian cows, but at least two pairs of eastern great egrets nested successfully along Barongarook Creek.

White-necked herons must have been nesting on an industrial scale from around the time of the first rains, and within 2 years the few locally raised youngsters that could be seen fishing in small family groups around the lake were joined by hundreds more, and dispersed more widely in large groups. On one occasion I found four flocks of between 20 and 40 young birds between Colac and home, a distance of just 30 kilometres, mostly standing around in the shallow fringes of dams though some were stalking through wet paddocks.

Although black ducks had always been present around the lake or in the larger creeks, as bird numbers increased they were joined by impressive armadas of chestnut teal, mixed with smaller numbers of grey teal. Whatever they were feeding on most certainly lived on the lake floor, and in midsummer there were

Young white-necked herons practise the difficult art of picking up sticks as they stalk through a flock of pelicans.

Chestnut teal feeding in the rising shallows.

sometimes thousands of dabbling duck bottoms visible at any one time, extending up to 30 metres from the shore in these shallow waters, and staying under for so long that only one head among a dozen or more birds would be seen above the surface at peak feeding times.

The white-edged wings of distant hardheads could also be seen among groups of flying ducks for a few weeks before they moved on, but other visitors were more difficult to see. Small groups of spoon-billed ducks that might have been shovelers could be seen a long way out, but it was only once these started appearing in small numbers among the dabbling ducks closer to shore that they could be recognised as zebra (pink-eared) ducks. Beautifully striped and with a curiously shaped, spatula-like beak designed for filter-feeding, they lingered for several weeks after the hardheads, but disappeared just as abruptly.

In 2009 a mixed flock of terns, including some Caspian terns with their slightly oversized hairdos, had briefly visited the lake as it began to refill, but early in 2012 flickering sparks of white began to appear far out over the lake, their numbers soon building up into thousands which would descend *en masse* to carpet the mudflats near Barongarook Creek. With their variable colours and patterns, they were probably mainly whiskered terns of various ages, but the conflicting opinions among the binocular-wielding humans wandering the shorelines suggest there may have been other species mixed in.

Sharp-tailed sandpipers also arrived in their thousands, wading through the shallows through a background of mudflats turned silver by the carpet of terns. Among the bewildering profusion other birds could also be observed in smaller numbers, from the handsome colours and contrasts of red-kneed dotterels to

Sandpipers forage among a resting flock of whiskered terns.

curlew sandpipers, with the bodies of sandpipers and the curved beaks of curlews. Deluged with an abundance of prey the scattered swamp harriers along the shoreline only needed to hunt for short periods, usually losing interest by mid-morning, and the great flocks of resting birds became quite blasé about the occasional raptor that drifted past, high overhead.

Rails and crakes are notoriously shy, and even though they must have been breeding successfully as the lake water level rose, until this time you would be lucky to catch anything more than their occasional sharp cries and groans from the dense sedge beds. With the arrival of the shorebirds and the greatly diminished threat from the harriers, buff-banded rails and spotted crakes suddenly came out of the woodwork in droves. If you sat still on the jetty near the botanic gardens you could sometimes watch half-a-dozen crakes foraging through the reeds a few metres away, and rails would skitter across your path wherever you walked near the water. Looking closely at some images of the massed terns, sandpipers and other abundant small birds, I could count up to 10 rails wading a long distance from the nearest cover, indifferent to any danger from above.

The hoary-headed grebes that began to appear in small flocks through the spring of 2011 also reached their greatest numbers by midsummer, by which stage they were perhaps the most common bird on the lake. On a 5 kilometre bike ride along the shoreline on an early summer day, I estimated between 500 and 1000 grebes per kilometre along the southern shore of the lake before the track became too soft to ride further. On still days they would form a snaking ribbon between 15

Hoary-headed grebes begin to appear in considerable numbers.

and 30 metres from shore, presumably following the greatest density of insects below, while on windy days they would drift in rafts of hundreds of birds that scattered widely over the lake.

Some of these birds may have been Australasian grebes, which are very similar when out of their breeding plumage, as these had bred in small numbers in the bird reserve in the previous summer, and they were joined by a pair of great crested grebes for a few weeks. As the summer came to an end small groups of Australian shelducks began to arrive, scattering through the already dwindling flocks of ducks. The more nomadic species had already been gone for weeks, though there were still many large groups of chestnut teal with the males still in their rich breeding colours. The terns and sandpipers had also mostly vanished apart from odd strays that lingered a few days longer near pairs of stilts.

In my records for the beginning of autumn I noted that nearly all of the chestnut and grey teal left, literally overnight, after 3 days of flooding rain further north. Though the floods were disastrous for humans from central Victoria to the Murrumbidgee and as far north as Sydney, for the birds it created the potential for a new season's breeding in freshly filled waters. Black ducks had thinned out months before though there were still small groups around, and nearly all coots, swamphens and stilts had also gone, though these could still be seen in moderate numbers around the northern shores of Lake Corangamite at times.

Black swans and shelducks were still moderately abundant along with flocks of increasingly morose-looking seagulls, but the lake looked strangely deserted after several months when waterbirds had been so abundant that it could be difficult to see the mudflats they were resting on. A few rafts and ribbons of grebes still drifted far out on the lake, not surprising given the continuing abundance of insect life in the shallows, but though other minute animals, including copepods, were still abundant the only larger animals feeding on them were the increasing numbers of fishes. Hoary-headed grebes are reputed to be flighty birds which take off readily if disturbed, but it was only now in the absence of surrounding flocks of other species that they began to show signs of restlessness, lifting off at the slightest disturbance as if eager to move on elsewhere.

A week later the shooting season began, reintroduced in the previous year by the recently elected state government even though surveys showed that the great majority of Victorians oppose the killing of wild birds. With the ducks already mostly gone the hunters can't have had much luck, but by the second morning of the season the last grebes were also gone, leaving only a few scattered and very nervous black ducks that panicked into the air if you went anywhere near them. The day remained grey and though the sun emerged fitfully late in the morning even the insects seemed to have fled, and it would be months before waterbirds returned again in moderate numbers.

Lake Corangamite

Eighteen thousand years ago Australia was much larger and drier than it is now. Sydney Harbour was a pleasant, wooded valley, and you couldn't even hear distant surf on most of the plain that would eventually become the Great Barrier Reef. Freshwater plants and animals moved freely between northern Australia and New Guinea, which were joined into a single landmass called Sahul, and most of the world's fresh water was locked away in the ice caps and glaciers of the Northern Hemisphere.

As the ice melted over the next few millennia sea levels rose around 130 metres, before dropping a little to the present level. Some droughts during this time span lasted centuries, and hiccups in the world's changing climate also caused occasional slips back to conditions more like the ice ages. If I could step back around ten thousand years to the time when what we like to describe as civilisation was just beginning, the world within a day's walk from the desk I am writing at would be a very different place.

The fast-moving, perennial streams flowing south from the ranges I can see over the top of my computer screen were probably dry gullies except after heavy rains; this is why the fishes now living in them all have a marine stage in their life cycle, though it doesn't explain why glass shrimps are also abundant in most of their estuaries. Where those dry gullies crossed the future thread of the Great Ocean Road, currently one of Australia's iconic tourist attractions, the sea was just a distant sheen to the south-west and a few centuries earlier you could still have walked and/or waded from my front door to Tasmania.

A few true freshwater fishes still survived in the inland-flowing Barwon River, which passed not much over a kilometre away to join the Yarra River out on the scrublands that would eventually become Bass Strait as the sea continued rising. In the often dry conditions of the time the Barwon may have been little more than a chain of ponds at times, but it must also have carried some great floods. We know this because there were still active volcanoes a few kilometres to the north, and a lava flow from one of them blocked the Barwon, diverting it west to form a huge body of fresh water.

With an area of over 250 square kilometres when full, Lake Corangamite is still by far Australia's largest 'permanent' lake, but it was seven times larger then. The lake is now just a shadow of what it was when I was first saw it, with much of its living diversity gone due to a combination of drought and ever-increasing regulation of its levels, intended to protect adjoining grazing lands. Yet it still has a certain grandeur with the wind whipping up salt spume as a squall moves across from the north, to heap mounds of tiny snail shells among the succulent red stems of glassworts.

Lake Corangamite, autumn 2011.

10
Predator and prey

Drought, flood and water quality will all shape the ecology of any wetland, but the interactions of the diverse groups of animals and plants that make their homes here are just as powerful in their effects – particularly the impacts of predators, which can be the most effective long-term force of all. Apart from regulating population size in the various prey animals they feed upon, predators have also helped to shape the evolutionary history, development and ultimate fate of entire groups of animals, creating an intricate mosaic of wetland types and species I think of as a blend of old and new ecologies.

Consider some of the more primitive crustaceans such as anostracans (fairy and brine shrimps), now found mainly either in ephemeral or highly saline waters, where more recently arrived groups such as fishes and dragonfly larvae are rare or are unlikely to survive long enough to become significant predators. Other primitive crustaceans have been pushed still further, for example the mountain shrimp and its close relatives which are now found only in high-altitude streams, caves and buttongrass plains in Tasmania, as well as many other relict species that have also been driven into the ecological backwaters of the world by the evolution of increasingly efficient predators.

Yet many other crustaceans such as copepods and water fleas remain spectacularly abundant in a diverse range of habitats, because they have adapted to predator pressure by breeding so prolifically that they have now become a lynchpin in these newer ecosystems. The largest crustaceans have taken a completely different tack, armouring themselves against attack and hiding in confined spaces while they are still soft after a moult; the body design of crayfishes, shrimps and crabs has literally been shaped by predation. It must be a successful strategy because these are among

A female musk duck presents a small yabby to her offspring.

the most abundant animals in diverse types of wetlands, even though they are a favoured food for a wide range of predators from fishes to birds and humans.

What it takes to be a predator

The animals we regard as true predators generally follow one of two major hunting strategies: creep-and-seize, or chase down by a superior burst of speed, though even the fastest predatory fishes rely on a certain degree of stealth to get close to their prey before making their attack. Water is hundreds of times denser than air, so streamlining is essential for any submerged predator that hunts at speed, and the power source which drives fast-moving fishes and waterbirds that hunt underwater is either the tail, or legs set well back along the body.

Most aquatic predators are cold-blooded so they don't have to feed as often as birds or mammals, and they can sit patiently through days, weeks, or sometimes even months of hunger, without any obvious evidence of distress. Insect larvae such as water tigers are particularly good at this, and move so little when there is no food available that algae and protozoans will grow over their unmoulted skins. Yet they are still willing and able to sneak up on impressively large prey when opportunity arises, increasing in size dramatically just days after their most recent, heavy feed.

Even fishes can adjust their activity and growth rates according to the amount of food available, and if I have to leave my aquariums in someone else's care while I'm on a field trip the fish aren't fed until I return, because even a small overfeeding could cause more harm in terms of toxic wastes than a few days without food. It doesn't take long for them to recover: among the most impressive famine-and-feast fishes I have kept are Pacific blue-eyes, which can swallow so many mosquito

Barramundi: a powerful and well-streamlined predatory fish with reflective eyes for night-hunting.

larvae in a few minutes that their backbones bend backwards from the effort of swimming; yet they will look as svelte and elegant as ever just a few hours later.

The balance between predator and prey is an interactive process, shifting back and forth as we have seen in earlier chapters, with other examples still to come. Extremes of drought and flood may kill off or drive predatory fishes, most insect larvae and even waterbirds to refuges that rarely dry out, and it takes time for the predators to return or re-establish in all of their old haunts. In the absence of any of their normal predators many smaller animals can multiply on a monumental scale, ideal conditions for allowing every possible combination of the genes available to be experimented with, with any failures to be winnowed out later.

Sooner or later every species will have to live with predators again, weeding out those that are just a little too slow or careless, and improving the fitness of the species as a whole in the limited Darwinian sense. It is a routine that has been practised and polished for millions of years, shaping the very nature of every type of wetland animal, and species that aren't constantly exposed to predators can develop some bad habits in evolutionary terms. One isolated population of the Eacham rainbowfish lived for so many generations in a lake with no larger fishes that it lost all notion of avoiding them, and disappeared within a short time of several such predators being introduced to the lake. Fortunately the same species remains reasonably common in some nearby rivers, where predatory fishes have always been present.

A white-faced heron takes some time off from fishing.

The one thing both predators and prey share a liking for is shelter, particularly in the form of what we call 'snags' in Australia. Nicknamed for the way they grab onto fishing hooks by people who probably had little idea of how important snags are to the survival of most of the fishes they regarded as prey, these allow smaller aquatic animals to build up in relative protection – and the larger their populations become, the easier it is for predators to make their living from the increasing numbers of careless individuals that stray too far from their shelter. And snags are probably the ideal place for a predatory waterbird to rest and groom itself, relaxed yet always aware that the waters below could provide its next meal at any time.

Hidden predators – arthropods

What is a predator, and what is prey? The concept of food webs often doesn't seem to mean all that much in wetlands; it's more a matter of what you can fit in your mouth at any particular time. Both mudeyes and water tigers (the larval stages of dytiscid diving beetles) are readily eaten by many species of fishes, yet the larger ones can themselves be significant predators on young fishes as well as other aquatic animals, taking on prey their own size and larger. But they aren't fussy about what they eat and will also feed on their own brothers and sisters without hesitation, as long as they can take them by surprise.

To understand the very different ways in which these two highly successful yet unrelated predators capture and eat their prey, we need to look more closely at the structure of the insect head, with the mouth actually inside while the jaws and so-called lips do their work on the outside. Primitive insects evolved from some kind of worm-like creature, presumably with some kind of rudimentary legs on most of its segments. As insects evolved, these turned into three well-defined blocks: the head, then the thorax which is made up of three segments each carrying a pair of legs plus the wings above, and the legless abdomen which is usually 10 or 11 segments.

What wasn't mentioned in chapter 2 is that the head of an insect was also originally made up of five segments, three of which once carried their own pairs of legs. These segments are now thoroughly fused and unrecognisable in an adult insect, though they can still be seen in insect embryos before these hatch. In adult insects the three pairs of legs on the head have become mouthparts: the first pair of jaws known as mandibles, which usually do the initial job of chopping and ripping up prey; a second pair of jaws (maxillae); and a lower 'lip' (labium). What passes for an upper lip is really just a reshaped part of the actual head, which combines with the lower lip to hold food in.

If you went blank part way through that explanation of how most insect mouths and jaws work together, it can be simplified as mandibles for initial chopping up, with food going into a pocket inside the hard lips, and the maxillae to do the final mincing within that pocket. Once chopped up and ready to process further, the food *then* enters the head and goes into the mouth, a multi-stage arrangement that explains how insects can directly process much tougher foods (including plants) than any worm can deal with.

The only problem with this simplified scenario is that some groups of insects have done some strange things with their mandibles, antennae and legs in the quest to create their own ideal carnivorous face, and among aquatic insect larvae there are almost as many exceptions as there are forms that comply with the basic ground rules. Two of the more dramatic exceptions are described here, but there are also other bizarre variations, such the antennae of phantom midges which are used to seize mosquito larvae, rather than for their original sensory functions.

Adult dragonflies and diving beetles have fairly normal chewing structures (for an insect, anyway), but their larvae (known as mudeyes in the case of dragonflies and water tigers for the larger waterbeetles) have some spectacular variations on this theme which make them particularly effective as ambush predators. To my mind these curious exceptions suggest that both of these larval types evolved at some ancient time while the head plan of insects was not yet completely decided, and the adults are partly a kind of afterthought that evolved as a better way for these dramatic carnivores to commute between wetlands.

In mudeyes the so-called lower lip isn't a part of the chewing assembly, but is a folded and hinged apparatus that can be extended at a blinding speed to seize prey with formidable talons, so that the mudeye doesn't need to risk frightening prey by creeping too close. This pair of talons looks remarkably like a pair of specialised seizing legs because that is exactly what it is, and has probably been for more than 300 million years. Remember that the primitive insect head was made up of several segments that are now fused, each with its own limbs attached though these are now long gone in most contemporary insects. It is humans who have made this structure seem hard to understand by renaming it a 'lower lip', because our vertebrate minds aren't comfortable with the idea of predatory, grasping legs growing out of the bottom of the head!

Once a mudeye's prey has been seized it is pulled back and gradually fed through the mandibles, which slice the victim up, alive or otherwise. Folded away, the lower lip/legs look like a duck's bill strapped on carelessly under the head, as can be clearly seen on the damselfly larva victim in the photo below. Close relatives of mudeyes and themselves formidable predators on smaller animals, even the smallest damselfly nymphs have a huge appetite for micro-life such as copepods. As a result it is a rare event to find significant numbers of these

Mudeye devouring a damselfly larva; the lower 'lips' on both animals can clearly be seen.

A water tiger captures a flathead gudgeon.

otherwise prolific crustaceans in any waters where damselflies have had a good breeding season.

A water tiger is also a stealthy hunter, but it must be virtually in contact with its prey before it attacks, and even then it seems somewhat uncoordinated compared to the gradual approach and precisely aimed lunge of a mudeye. Instead, its flattened, very flexible body swivels at a snail's pace as it senses the approach of a potential victim, sometimes twisting almost through a semi-circle while it decides what to do next. If the cause of the disturbance seems too large to eat, it may be a potential predator, in which case the water tiger waits until it is gone before it moves again.

However, anything in the right size range will be abruptly seized once the larva's jaws are in position, just a few millimetres away. The biting apparatus in this case is the mandibles themselves, in water tigers looking like a pair of sickles that can be driven into the body of the victim. As these have been modified so they can no longer chew or shred, the water tiger processes its prey by injecting it (via the mandibles themselves) with enzymes that turn its innards to a liquid that can then be sucked back into the larva. It seems likely that the injected enzymes also contain some kind of anaesthetic, as the victim usually doesn't struggle much. Even a fairly large fish freezes with its fins erect almost immediately after being captured and, apart from the occasional spasm during the first few minutes, puts up no real resistance; over a few hours the victim blackens and shrinks as its tissues dissolve and are absorbed.

Spiders are more distant relatives of the insects, but apply a similar approach to water tigers for digesting and sucking out the contents of their victims. Of all the arthropod predators they are generally much less common in wetlands than their more successful insect cousins, with the exception of long-jawed spiders, which specialise in damselflies so their webs may sometimes form a continuous drapery along sedges and rushes fringing ponds and lakes.

A slender male long-jawed spider courting a female in her damselfly-wing littered web.

And carnivorous water bugs have created their own distinctive variation on the water tiger's technique, with their mandibles fused to become a single piercing and sucking beak. It is likely that these bugs also inject a paralysing substance along with digestive fluids, but with the mouthparts reduced to something like a straw it is their front legs that are used to seize and manipulate the victim.

Fishes as predators and prey

Compared to the elaborate head structures and long evolutionary history of the most successful insect predators, the approach of most of the supposedly

Olive perchlets, one of the most widespread of the glassfishes.

more-evolved fishes can better be described as snatch-and-grab. Most larger predatory fishes are solitary hunters, whether they actively chase their prey or strike from an ambush, while smaller fishes which are often a favoured food for their larger kin may form schools to reduce the odds of any one individual being caught. Schooling fishes often have distinctive patterns or stripes that help keep them together by day, though at night they scatter because they can no longer see.

Rainbowfishes are among the most abundant schooling fishes in the northern half of Australia, and in some rivers they may be so dense that it becomes impossible to see more than a few centimetres through them even in the clearest waters, so fast and furious is the action as thousands of these fishes pour past your facemask. They are often joined by glassfishes to form loose schools for mutual protection, a visually improbable combination with the semi-transparent glassfish hopping and jiving while a gaudy stream of rainbowfish streams through them.

Schooling is an effective distracting strategy, to the point where some of the larger predatory fishes hunt by night when the schools break up, and the reflective eyes of the night hunters create an eerie impression if you are snorkelling by torchlight. One of the largest of these is the barramundi, which lurks under vegetation and in dark places during the day, emerging as darkness falls. Although its eyes are small in proportion to the size of its body, the reflective layer within the eye is so precisely arranged to collect every trace of light that torch reflections from the eye appear as a shaft of light through slightly murky water, with a spot of bright light marking where the fish is looking.

Fishes are also the main predators on many types of aquatic insects and larvae, but apart from waterboatmen and backswimmers they are not generally keen on water bugs as food. Some of these probably don't taste very good, and many terrestrial bugs smell like an industrial-grade version of coriander, a pungent herb named after the Greek word for bug: *koris*. But there are also many waters where fishes aren't found, and it is waterbirds that are the most widespread and adaptable wetland predators.

Waterbirds as predators

Predatory waterbirds are far more intelligent than most of their prey and, combined with their ability to fly, this has allowed them to develop a greater diversity of hunting strategies than any other major group of animals. Few predatory waterbirds are found together in large numbers except when feeding is easy and easily accessible, such as in a drying waterhole, but these temporary aggregations disband and fly off in all directions once the food source literally dries up. There are some exceptions such as pelicans which are bonded by their communal fishing habits, and cattle egrets which forage in large flocks among cattle, though not usually in wetlands except as isolated individuals.

Herons and their relatives are long-legged hunters that combine a slow stalking pace with a last-minute lunge that is much faster than any underwater predator can achieve, because this type of strike builds up most of its momentum in air, and the streamlined beak can drive deep into the water before it is slowed down significantly. Kingfishers are much smaller so they take advantage of the low density of air to achieve comparable accelerations, but because of their relatively small body size their entire body must be the missile, and their prey must be close to the

A short section of Turn Back Jimmy Creek with around 30 herons and egrets, and 10 pelicans feeding on fishes trapped in its evaporating waters.

Eastern great egret in typical stalking pose.

surface as they don't have any of the swimming ability of birds that usually hunt underwater.

The flattened feet of predatory birds such as cormorants and darters that swim beneath the surface are arranged so that the webs of their feet fold together as they are brought forward, only opening out on the reverse stroke. For maximum power

Australasian darter.

delivery their legs are set so far back under their bodies that they have some difficulty in walking, and as their flight ability isn't all that impressive and their wings waterlog easily, cormorants need to spend long periods of time drying their outspread wings between feeding sessions. These are one of the few predatory waterbirds often found in groups while resting, but they will also hunt cooperatively in flocks at times.

Darters are one of the most peculiar birds imaginable, a novel blend of elegance and awkwardness, with an elongated and elastic neck designed for seizing fishes as they wing their way underwater that gives them their alternate name of snakebird. Even more aquatic than cormorants, darters often swim with their body largely submerged, looking markedly reptilian until they emerge from the water. With their large and somewhat flaccid-looking webbed feet and craning neck movements, darters on land look more like cartoon characters than sinuous and highly effective hunters.

Crocodiles

In the grand scale of things most reptiles are relatively minor predators on wetland animals, though file snakes are common enough in tropical floodplains that they must take a fair toll of rainbowfishes and similar small species. It is crocodiles that

A 50-million-year-old crocodile skull from Gladstone, Queensland.

dominate the wetlands of the far north, as they have for many millions of years. The smaller freshwater crocodile is a relatively innocuous species that feeds on a range of unglamorous prey from insects and crustaceans to fishes; rainbowfishes are apparently a particular favourite. The saltwater crocodile is an altogether larger and meaner creature, noted for its occasional attacks on humans, and making exploration of any of its diverse wetland habitats a nerve-wracking procedure.

Although these are lethargic beasts that spend much of their time sunbathing, they are also very effective ambush predators, and can sprint briefly on land faster than a human can run. Juvenile 'salties' may simply wait in shallow water for fish or shrimps to pass within snatching range, but will also rush at prey if hungry. Adults often stalk potential prey such as waterbirds, turtles and mammals, approaching the shoreline underwater and lunging out of the water once in range. Abundant before the recent era of intensive hunting, saltwater crocodiles are now protected and have recovered dramatically in numbers, returning to their former position as the top predator of tropical rivers and estuaries.

Human impacts: hunting, fishing and aquaculture

A dead marron washed up near the mouth of an otherwise apparently pristine river on Kangaroo Island hints at the hidden changes introduced crayfishes have caused below the surface.

Humans were relatively minor predators in wetlands long before European settlement, but it is unlikely that Aboriginal peoples had significant impacts in these environments, and even birds such as magpie geese, whose eggs were harvested on a considerable scale, were common over much of Australia until the advent of the gun. Hunting continues to be a major destructive force on wild birds worldwide, and though Australia is much less affected than European countries where even the tiniest birds may be shot indiscriminately, there are still probably tens of thousands of waterbirds injured on the opening day of every shooting season.

I don't hunt, but have some sympathy for those who do so responsibly and, as I also like to fish in a small way it would be hypocrisy for me to condemn other hunters on phylogenetic grounds. Although it was commercial fishing combined with the increasing control of many rivers that made the first major impacts on indigenous fishes and other aquatic animals, recreational fisheries continue to cause incidental damage through carelessly discarded lines and hooks, and poorly envisaged or enforced laws on catching slow-growing animals such as many of the more localised freshwater crayfishes.

Other problems caused by introduction of alien fishes such as trout for recreational purposes are well known, but as these are now thoroughly naturalised it is too late to correct these on any significant scale. Goldfish have been spread widely through the aquarium trade, with carp illegally introduced as a potential aquaculture animal. And even though aquaculture is now fairly well regulated and is largely restricted to so-called native species, many of these have already escaped to become feral in places well outside their original range. The three main cultured and relatively adaptable crayfishes are a particular problem, with redclaw now established in some central Australian rivers, while yabbies and marron have gone viral since their introduction to Kangaroo Island, changing the underwater ecology there forever.

Part 3

Ecology: fitting into your environment

Fresh water falls from the sky as rain, and when it reaches the earth it either soaks in or flows downhill, changing as it moves through diverse landscapes but always following the path of least resistance. These places where water flows or pools, rises to the surface or disappears underground are all wetlands, and though it may not be obvious to the casual observer they are as diverse as any terrestrial landscapes, with very different combinations of plants, animals and water conditions. In the chapters that follow we start by looking at the natural pathways of flow, what these do to the waters that flow through them, and what it takes to be able to thrive in these varied conditions.

In later chapters we will follow the general paths of other waters through our ancient landscape, from clear and fast-moving streams as they drop from higher altitudes to their sluggish and muddy lowland phases; the marshes and swamps, backwaters and billabongs, and even lakes where they detour or sometimes just lose their way; and ultimately the complex meshwork of lagoons, estuaries and other tidal waters where inland waters join with the sea. Finally we look at the most recent and artificial wetlands created by human hands, poor in species and diversity but still potential habitats that could be used to repair some of the damage we have caused over the past two centuries.

Although all of the following chapters group wetlands together in particular ways to link their stories into coherent themes, keep in mind that there are no clear-cut boundaries between the different types of wetland other than those we draw in our own imaginations. Marshes may drain so far underground that estuarine fishes appear in them from the caves below, rivers dammed by landslips become lakes, salt lakes may form metres away from the overflow of a freshwater spring, and waterfalls may spill from coastal cliffs straight into the sea. The only thing that links all wetlands is water.

11

High and low places

Rain is virtually pure fresh water with almost nothing dissolved in it, and dissolved salts in the glacial lakes near the summit of Mount Kosciuszko have been measured at somewhere around 3 milligrams per litre – barely more than in distilled water. As the rain falls it collects a little oxygen plus a touch of carbon dioxide from the air around it, forming very weak carbonic acid which is quickly neutralised by minerals in the soils it comes in contact with.

The process of absorbing salts and other substances begins from the moment water comes in contact with the ground, slowly at first if it runs over rocks that have already been well weathered, but accelerating if it travels underground for any length of time. Water also absorbs dissolved organic matter such as tannins from peat, bark and leaves, tinting it to the colour of tea (see chapter 16), as well as collecting various other organic substances from the animals and plants that live and die around its course.

In the high gullies and ranges where water begins its downhill journey it creeps underground at first, until the many hidden seeps and soaks begin to surface and run together into small streams. Sometimes water rises by capillary action, through crowded grains of sand and spaces so narrow that the surface tension of the water is stronger than gravity; springs of this kind may erupt in unexpected places. But even before the water reaches the surface there are animals underground that are already reliant on these permanently wet places.

Water flowing underground

Some of the animals that live in the spaces between soil particles in wet places are not much different from those that live in a garden; there are worms, beetles and

Land yabby chimneys.

leaf hoppers everywhere in healthy soils rich in organic matter. The presence of permanent water not far from the surface, where the soil becomes waterlogged, also attracts a variety of aquatic animals. Most of these are very small, including some unusual groups of crustaceans and wheel animalcules, but burrowing animals can grow larger if they create their own spaces, in turn creating new types of habitat for other animals.

Land crayfishes (in south-eastern Australia known as land yabbies) excavate deep, multi-forked burrows into sodden soils, shovelling the excavated mud out to form an often-blackened chimney that looks like a small, dark volcano. Somewhere below the water's surface the excavator lurks, eating both living and decaying plant matter and probably the occasional worm or any other animal matter it can scavenge, and sharing its burrow with amphipods and other animals that move too quickly for it to catch in these confined spaces.

In these highly specialised species the basic crayfish body plan has undergone some significant changes, with the tail (no longer needed as an emergency device for flipping the crayfish backwards) often shrunken to a disproportionately small size. Perhaps it would have disappeared already in some of the most profoundly adapted burrowers, except that it is still needed by the females to wrap around their eggs. The claws of many land crayfishes have also become strongly asymmetrical, with one broad but flattened claw probably used as a type of bulldozer blade, to push mud upwards and out of their burrows.

Although this impressive development may remind us of the signalling claws of the marine and estuarine fiddler crabs, it is unlikely that land crayfishes use it for this purpose as they are rarely seen above ground during the day. In any case this larger claw is no more brightly coloured than the other one, which is often much more slender and covered with a veritable forest of fine bristles that must have some function as sensory hairs in the dark waters below.

Land yabby showing the relatively small tail and exaggerated claws.

What they are used to detect we don't know, as these elusive animals are difficult to catch, let alone study in their deep lairs where even the water is too murky to allow photography. I've experimented with a self-lighting plumber's camera on a 2 metre cable, and its journey downwards into the increasing darkness of a burrow, projected onto a hand-held computer screen can best be summarised as 'getting muddier … darker now … muddier still …'. Even now, we still don't know how many species of land crayfishes there are, and as they don't seem to be great travellers there may be dozens more still to be recognised.

Land yabbies may also be found in heathlands and damplands, poorly drained places where a combination of acidic, nutrient-poor soils and the accumulation of partly decayed vegetation known as peat encourage a distinctive flora. Here, the water table rises during wet periods until the soil may quake like a sponge, forming pools in places, and although few animals apart from insects live here the spongy peat acts as an underground reservoir, releasing water slowly to keep streams flowing even through the dry season.

Carnivorous plants

Plants that feed on smaller animals have evolved in many different ways, and from unrelated families. What they all have in common is living in nutrient-poor environments such as peaty swamps, where nitrogen, potassium and phosphorus

Pink bladderwort.

are usually lacking. On the other hand these nutrients are abundant in animals, and carnivorous plants have evolved many curious ways of trapping the animals and extracting what the soil lacks to fuel their growth. Although technically carnivorous, all of these plants must still photosynthesise to grow, and their elaborate traps are usually just highly modified leaves.

Bladderworts have tiny bladder-like traps that suck minute animals in; the bladders on terrestrial species are underground, but they are much more conspicuous on the free-floating, rootless aquatic species. The leaves of sundews are covered in sticky droplets that give them their name, and these not only collect small insects but may even help to digest them. There are two families of pitcherplants in Australia, the northern species growing in trees rather than in wetlands, but the Albany pitcherplant in the south-west has small, elaborately configured pitchers that trap mainly ants. The maggots of a highly specialised wingless fly live in the pitcher, feeding on decaying insects, and for some bizarre reason these adult flies mimic the species of ant that are the pitcherplant's main prey!

Albany pitcherplant.

But even streams may vanish underground at times, leaving animals trapped in pools that grow ever shallower as the water recedes. A half-kilometre section of the nearby Barwon River disappears under the rocky rubble that lines its bed late every spring, and though I try to rescue river blackfish and galaxiids from several deeper pools where they are regularly stranded, the timing is always a bit hit-and-miss as the water table may suddenly drop close to a metre overnight if there hasn't been rain for several weeks.

River blackfish and galaxias trapped in a pool of a river that is sinking underground.

Water that sinks into the ground may become groundwater, permanently flowing deep below or pooling in porous rocky basins when it can go no further, but it may also resurface as springs at lower altitudes. Water quality in springs is variable, depending on what it has passed through, so springs that are fed by peaty heathlands are usually dark, acid and fairly sterile in biological terms. Springs that emerge from limestone are usually clear, and while their water is hard due to dissolved calcium and magnesium salts, they are not usually particularly saline as limestone doesn't contain much in the way of sodium salts.

South Australia has some of the most extensive limestone springs and submerged cave systems in this country, and they look from above like many other wetlands in the area, but open out into spectacular vistas below. The waters here are comparable in clarity to the Great Barrier Reef, and you can see close to a hundred metres to the far side of the first pond at Ewens Ponds when you duck your head under. Even after dozens of dives here over the past 50 years I still find that first moment of snorkelling here literally breathtaking, looking across continuous carpets of vivid green plants that grow as much as 6 metres deep. Drifting down the slow-flowing raceways that link the three ponds you are surrounded by schools of galaxiids and pygmy perch, while spiny crayfishes wander around the fringes of the springs near the bottom of the deeper pools, where fine gravel seems to boil in the deepest hollows.

Near the edge of the chasm in Piccaninnie Ponds.

Piccaninnie Ponds is even more dramatic, from above looking like a shallow marsh broken up into an irregular mosaic of pools by stands of water-ribbons, but as you swim through these a deep, blue chasm suddenly opens up, apparently glowing with a spectral light from below, and disappearing far down into the stone. Freshwater turtles are common here and are easy to observe as they forage through the shallows, but even more striking is the array of marine and estuarine fishes that come and go freely from the sea via underwater tunnels. Yet even though the sea is just several hundred metres away, the springs flow strongly enough that there is little trace of salt in these crystalline waters.

Predatory waterbirds aren't abundant in these clear springs despite the large numbers of fishes that can be present, partly because the fishes are mostly too deep for convenient hunting, and because they can see any predator approaching from a long distance. But that same visibility will allow a snorkeller to watch cormorants hunting, or coots diving to pick the prolific caddisfly larvae off the massed plants metres down from a distance that doesn't seem to worry the birds.

Where the salt goes

The path followed by water on its way to the sea can be tortuous as it dips underground, runs in streams, or spreads out widely through marshes and lakes, until it joins some of the most complex and productive ecosystems in the world: estuaries, which are looked at in chapter 17. In the lower relief landscapes of inland Australia water flows much more slowly, giving it time to absorb more salts, and as the landscape is so flat it often doesn't get anywhere near the sea. Lake Eyre, for example, is below sea level, but it doesn't take much in the way of higher ground to trap water in smaller lakes and watercourses even above sea level, where evaporation concentrates the salt year by year.

Drifts of salt-lake snail shells along the shores of Lake Corangamite.

The water in salt lakes is also often rich in calcium and magnesium, giving it a somewhat bitter taste, but it is sodium salts that are usually the dominant ingredient and they create a very different type of wetland to any freshwater ecosystem. Only a few species can tolerate the extremes of temperature here, along with wild fluctuations in salinity from one year to the next, when fresh flood waters from monsoon rains may carry young fishes, shrimps and other animals down from the north. Although these may thrive for a time and can even grow to substantial sizes, there is little chance that they will breed here and most of them are doomed as evaporation takes its toll, with the lakes they are trapped in becoming increasingly saline.

Animals that thrive in salt lakes and other alkaline inland waters aren't particularly diverse, but in the absence of any predators worth worrying about they can become extremely abundant. Under the right conditions salt-lake snails can form a living carpet across the lake floor, and heaped windrows of their dead shells may pile up for many kilometres along the shoreline. Crustaceans are the most common planktonic animals in the waters above and include some fairly large species such as the giant pearl-ostracod, a creature whose beautiful hues remind me of the glossy, pastel shades of some freshwater pearls.

Native brine shrimps are also common, though in some coastal areas they have been replaced by an introduced species used as an aquaculture food. The males of these transparent animals develop a fleshy-looking moustache which is used to clasp the female during mating. Like many of their close freshwater relatives the fairy shrimps, the tails of native brine shrimps are tipped with a curious,

Giant pearl-ostracods, unusually large seed shrimps.

Native brine shrimp males.

bright-orange fork. Presumably they must be able to see and recognise this unless its sole purpose is to attract predatory waterbirds, as it does a fine job of spoiling the camouflage on an animal that is otherwise almost invisible.

As in many other waters, copepods are usually the most abundant planktonic crustaceans, and are often coloured red or orange by the algae they feed upon. Too small to attract waterbirds and with no other predators worth mentioning, the females of some salt lake species rarely carry more than four or five eggs, though these are much larger in proportion to the size of the female than those of fresher waters where dozens of eggs must be produced to allow for casualties. As the female probably doesn't live to breed again, this suggests that half of her very large offspring are likely to survive to breed in their turn, a clear indication of the advantages accruing to animals that are prepared to adapt to less popular locations and conditions.

On the minus side, the salinity can vary so much over the course of a year that completely different suites of crustaceans will become the most abundant animals at times. The lakes may also dry out at times, though this isn't much of a problem for animals that developed hard-shelled eggs hundreds of millions of years ago, and which may now rely on their very few predators to carry the dormant eggs to new habitats that should suit them.

Black-winged stilts and red-necked avocets rest in typical one-legged pose.

Not surprisingly their main predators are birds, the bioregional equivalent of filter-feeding flamingoes on other continents. Like flamingoes, stilts and avocets walk tall on slender legs, and feed on small planktonic animals in waters that are often very saline, but they have their own eccentricities, including incongruous calls such as yelps, toots and other less easily described sounds. Despite their slender and feeble legs these birds are good swimmers, usually wading while they feed but looking more like flocks of peculiar-looking ducks out on deeper waters.

While stilts are most likely to be seen wandering through shallow waters, red-necked avocets seem more versatile in their approaches to feeding. Groups may be seen foraging in tight formation near the shoreline, apparently driving some kind of small animals ahead of them, and closing in on concentrations of their prey at times. But they will also form loose flocks well out from shore, swimming and diving in the short, sharp chop that even the lightest winds build up across open lakes. Presumably the hunting is good at such times as their distant yarping carries even to the shore, suggesting excitement as the bobbing avocets call only during the brief moments when their elegantly curved beaks lift above water.

Water in the desert regions doesn't just fall from the sky or flow southwards during extreme Wet conditions in the north; some of it returns from far below where it has been brewing for millions of years. Although the Great Artesian Basin that underlies much of central Australia has been bored and pumped to bring its often hard and saline waters to the surface, it has also been making its way upwards through springs in many low-lying areas for thousands of years. The alkaline minerals in the water form mounds as the spill runs over their sides, gradually raising the spring above the level of the surrounding land, and although these tiny wetlands are just minute blips in a vast and arid landscape they can be surprisingly rich in distinctive animal species.

As usual there are unusual, relict populations of crustaceans, but there are also many specialised fishes that thrive in these long-isolated waters, including blue-eyes, hardyheads, eel-tailed catfishes and gudgeons that must have arrived during much wetter times, as they have evolved into distinct new lineages. Among all of these fishes a small group of five blunt-headed gobies, close relatives of a marine species that still lives in saltmarshes and other estuarine environments of the far north, are exceptionally well adapted to desert conditions.

The inland species have made themselves at home in isolated springs of bitter and often very saline water, but the most common species has also colonised bores drilled to provide water for stock. Not surprisingly, the desert goby can tolerate salinity ranges from virtually pure water to nearly twice that of sea water, in waters that may be alkaline enough to blister human skin. Although water doesn't fluctuate as wildly in temperature as the air above it, the desert goby must be able to tolerate lows of 5°C, and during hot spells when even the water temperatures reach over 40°C it avoids suffocation by clinging with its head above water, keeping its gills just wet enough to breathe air.

Male desert goby.

Troglodytes

Most underground water in Australia flows through porous rock, often deep underground, and we know little about the animals that may live below except for those that occasionally drift through caves and sinkholes that open out to the surface. Yet there is probably a surprising diversity of life in these dark waters, and new communities of unnamed species are still being discovered, including some specialised waterbeetles several years ago. Blind shrimps are known from caves in northern Australia, and one of these is found with the blind gudgeons and cave eels of the North West Range, straddling the Tropic of Capricorn in Western Australia.

Limestone sinkholes in the North West Range.

The rainfall in this area is unpredictable though always low, with just enough soaking into the limestone hills through sinkholes and crevices among the spiny porcupine grass to support an intricate and virtually unexplored web of tunnels filled with clear, alkaline water. As the rivers of the region dried out over millions of years the fishes and shrimps retreated underground, eventually abandoning their eyes and nearly all trace of colour. A vestigial network of blood vessels is all that remains of the eyes of cave gudgeons, creating a pink blush on their otherwise silver-white bodies. Even more curious is the recently discovered Barrow cave gudgeon, a freshwater fish from an offshore island with no fresh water on the surface, known only from specimens found in deep bores left by drills.

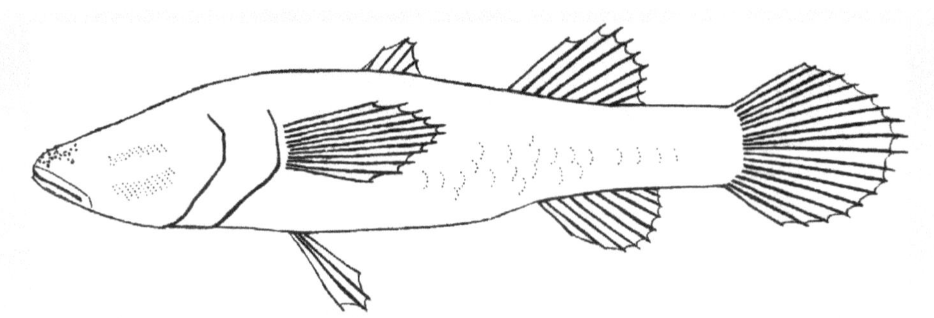

Cave gudgeon.

12

Streams, brooks and fast rivers

As water seeps out from springs, heathlands and other underground places it forms into streams which may be seasonal or perennial, but in their young phase are usually clear and relatively cool. In the Top End many streams and even smaller rivers run mainly during the wet season, becoming shallower and drying out once the rains have come to an end, but along the steeper slopes of eastern Queensland most rivers run through the whole year. In the south it is winter and spring rains that feed rivers and streams, but whatever the climate or latitude young streams are relatively sterile, with little in the way of food for the animals that live here except from what falls in from the surrounding forests.

These rivulets are usually known as creeks in the east, and brooks in the west. Regardless of the name, all such streams pick up minerals, sediments and organic matter as they cut their way through soil, or run over rock. Coming together into rivers, they slow when they reach coastal plains and gradually drop the heavier sediments to form a silty bed. During floods, slower-moving rivers spill over their banks across their floodplains, carving new courses through the sediments they have been laying down for thousands of years. Though linked across time and space, all of these phases create very different living conditions for both plants and animals. This chapter focuses on life in faster-moving streams and young rivers, while the next two chapters look at other aspects and types of rivers, including sluggish inland flows as well as the billabongs and other backwaters they create.

A shallow creek in the Northern Territory, crowded with annual plants which will soon be flowering and setting seed as the waters disappear.

On higher ground

Whether a stream starts in an open sphagnum bog somewhere above the snowline or surfaces among ferns and forests, one of the most striking aspects of its clear and rushing waters is the almost total absence of waterbirds. The high country is too high for most of these, and forest is far too dense for birds of open waters, but in any case there is very little food available for them here. The streams may be clear but there are few plants for herbivores, and though there may be plenty of smaller animals and often even some fishes these are mostly secretive, hiding among rocks and stones at the slightest disturbance.

The only animals present are those that can climb or fly: frogs and some lizards, insect larvae in the rocky rubble and occasional dragonflies or damselflies. Fishes are rarely present in the higher reaches because with very few exceptions they can't negotiate waterfalls, so even a short, vertical drop of a metre or so is enough to stop their upstream progress. Young eels can climb some wet surfaces, clinging to the rough rock face at the sides of a waterfall, or burrow through fine gravel to get past faster cascades, while juvenile lampreys may creep around the barrier over damp ground. Some galaxiids may be able to jump low barriers to their progress, but the climbing galaxias is the only member of the family that is regularly found beyond any substantial barriers.

Waterfalls can be formidable barriers even for fishes that can climb.

Staying put in moving waters

Most edible things wash away rapidly in faster-moving streams, and it is only a continuous supply of leaves and other organic matter falling from the surrounding forests and ferns that make it worthwhile living here at all. These add just enough dissolved matter to make it possible for algae to grow, supporting a diverse community made up mostly of insect larvae that remain hidden under stones during the day. Although the majority of these are herbivores that feed on algae or scavengers that feed on any organic materials they can find, a few species of specialised hunters also make their living here. Some of these larvae may spend most of their lives under a single large stone, until it is time for them to turn into the short-lived flying adults that will carry their genes to further waters.

Even at the best of times the water moves fast here, and the larvae have worked out many different ways to keep themselves from being washed away. Some glue themselves to the underside of rocks, using various types of sticky traps to catch tiny animals and other organic matter, so their food must come to them. There are also many more active larvae that crawl around on powerful, recurved legs designed to keep their flattened bodies gripped to the stone, with strongly jawed predators such as eustheniid stonefly larvae in pursuit of the ever-abundant mayfly larvae that graze on algae. But even these extreme adaptations are of little help during floods when the rocks themselves are thrown and tumbled around, and after heavy rains the rocks may seem strangely bereft of life when you turn them over, compared to the swarming hordes that are usually present.

Eustheniid stonefly larvae.

Mayfly larvae feeding on algae beneath a rock.

Spiny crayfishes

In terms of numbers and biomass, insect larvae are the dominant animals in the upper reaches of most streams, and the few crustaceans found here are mostly small and not very common. Yet paradoxically, those few crustaceans that may be present in even the tiniest trickle include many of the largest freshwater invertebrates in the world: the spiny crayfishes. These are mostly found from the higher-altitude streams of southern Queensland to the coastal streams of Tasmania, though one heat-tolerant species is widespread in the sluggish waters of the Murray–Darling river system.

On the coastal side of the ranges that run north and south along the eastern coast, many of these rivers probably joined on the continental shelf 14 000 years ago and at many other times in the deeper past. In the cooler conditions of the time, it is likely that there were spiny crayfishes passing unhindered between many river systems. As temperatures and sea levels rose, the crayfishes (which mostly have a preference for relatively cool, moving waters) retreated upstream, developing further into a diverse range of localised colour forms and species.

It sometimes seems that every tributary of every eastern river has its own distinct variety, varying dramatically in colour even though they may be genetically very similar. The southern Victorian spiny crayfishes I used to catch 30 years ago in the Yarra River, from near where this species was first described, are big beasts 30 centimetres and longer, and came in two very different blue-and-white versus green-with-red forms. In the nearby Barwon River the same species is only a third of that length, and is always a dark green with some bold red-and-white patterning.

Southern Victorian spiny crayfish.

This considerable difference between populations only 150 kilometres apart isn't necessarily dictated by the size of the stream they live in, because in Tasmania the giant crayfish known locally as freshwater 'lobster' is often found in the deeper pools of small, shallow and peaty streams that make even the narrow upper Barwon look like a respectable river. This is by far the largest freshwater invertebrate in the world, though it has some serious competitors among mainland spiny crayfishes, and is one of an island group of species that are regarded as different enough to warrant their own genus.

Slow-growing and even slower to mature, the so-called lobster has been overfished to the point where it is now completely protected, but the giant animals of the past are now just a memory and I have never seen anything larger than a 15 centimetre baby. How large these giants actually grew in the past is far from clear, and some official websites take the position that it wasn't necessarily all that much larger than the 5 kilogram monsters that are still occasionally found, suggesting a maximum weight *possibly* as high as 8 kilograms.

Other sources throw much larger figures around liberally, without bothering to offer any concrete evidence, and I have read several independent accounts of an old black-and-white photo that was supposedly hung on the wall of a pub in northern Tasmania. In these accounts the claws of the crayfish are always up at the shoulder height of the man who holds it and, depending on who you want to believe, the tail hangs at crotch level (credible), or reaches his knees (at something like 15 kilograms, starting to seem a little doubtful), or down at ankle level (an impressive 40 kilograms or more).

The most mysterious thing about this photo is that on an island where thousands of images of thylacines survive even though the animals themselves almost certainly don't, no one seems to have bothered keeping even a single photo of this crustacean King Kong. There are also biological reasons to believe that the most exaggerated accounts aren't possible, particularly given the immense size of the claws that can be seen on some contemporary images that are undoubtedly genuine.

The older any crayfish is and the heavier its claws become in proportion to its size, the more difficult it is to raise them in self-defence, or for any other reason. Eventually, it would become impossible for a really gigantic lobster to move at all except backwards, dragging its claws behind it, and natural selection must have put an end to any such ridiculous development long ago. Given that their primary diet is rotting logs and the microorganisms that break them down, any oversized specimen would probably starve to death once its cumbersome claws prevented it from getting at its low-nutrient food efficiently.

Apart from the considerable size that many native spiny crayfishes definitely do reach, they are also an impressively ancient lineage. Fossil claws 105 million years old from Dinosaur Cove, around 30 kilometres to the south-west of my writing desk, look very similar to those of some living species, including such details as the overall shape, the position and angle of the spines, and the heft of the hook. These ancient animals probably regard us as a passing nuisance in the game of life and probably expect us to go away eventually, just as the polar dinosaurs that once used to muddy their streams did long ago.

Fishes and fast streams

Somewhere below the last waterfall or fast-moving cataract, the world of most freshwater fishes begins, and in these increasingly wider and more open waters predatory waterbirds such as kingfishers and smaller herons return. The ecology of fast-moving streams in the northern half of Australia has probably not changed all that much since European settlement, mainly because there are relatively few introduced species in these waters – though, ironically, some 'natives' from other areas have begun to appear in places outside their natural range as a result of aquaculture and aquarium escapes in Queensland.

Rainbowfishes and glassfishes remain abundant here, and are still significant food sources for diverse predators from larger fishes to birds, and even freshwater crocodiles. The two families of catfishes are also widespread, and though the fork-tailed species are more likely to be found closer to the coasts some eel-tailed catfishes have moved a long way inland, though they usually avoid the fastest flowing sections of rivers.

Grunters are the most unfortunately named of all native fishes, a group that is more diverse in northern Australia than anywhere else in the world, yet the whole of this family has been saddled with a name based on the vulgar sounds made by

Coal grunter.

some species when they have been pulled from the water against their will. Despite the impression given by their common name most grunters are attractive and intelligent and, though these qualities are also associated with aggression and territoriality, that hasn't stopped some similar-looking cichlids from becoming among the most popular aquarium fishes in the world.

Perhaps it is time to rethink the grunter paradigm and create a name that more actively reflects their ecological significance as predators, as well as their good looks. The aquaculture industry has already made the first steps in this direction for obvious reasons, converting the crude and barbarous Barcoo grunter (see the next chapter) into the magical Jade perch, though the only obvious difference between wild and captive populations is that those that have been raised for food production usually look a bit on the pudgy side. Most grunters are fishes of murky, sluggish rivers, but a few such as the coal grunter have a marked preference for clear, fast-flowing waters, and these species are more likely to be boldly patterned and coloured than their cousins from murkier places.

The cooler, faster-moving streams of the south have changed much more dramatically than the streams of the north, mainly because of the widespread introduction of predatory alien fishes, particularly two species of trout and redfin perch. Despite this direct competition river blackfish remain reasonably common in many parts of south-eastern Australia, probably in part because they seem more

The ecology of many southern streams has been changed forever by the introduction of brown trout.

River blackfish.

tolerant of warmer and drier conditions than trout, and if temporarily trapped in still waters will find their way in among rocks where temperatures are lower than in direct sun. Unusual soft-skinned predators that live and breed among rocks and submerged tree hollows, male blackfish guard the eggs until these hatch and the young can scatter elsewhere.

Somewhere around where the faster-flowing streams start to slow down and consider maturing into rivers, pygmy perch begin to appear. These deep-bodied, wide-mouthed fishes look a little like some much larger carnivorous fishes such as the mouth almighty, but are constructed on a much smaller scale. Most of them are found in fairly clear and permanent waters in southern Australia, with a single and endangered species in the wallum country further north. Although often so abundant that hundreds can be caught in a single scoop of a hand-held net, pygmy perch don't form schools and they can be both significant predators on aquatic insects, larvae and smaller crustaceans, and important food items for larger fishes.

Other southern fishes such as grayling have probably become much less common as a result of clearing of land along rivers, along with the associated problems of siltation and deteriorating water quality. A silvery, elongated species that smells inappropriately like a cucumber when freshly caught, it was thought to be endangered as its sibling species in New Zealand died out many decades ago. In more recent years grayling seem to be making a comeback in less disturbed streams, or perhaps it is just that more people have been looking for

Galaxiids: cryptic diversity in the south

The galaxiids (pronounced 'ga-lax-ee-ids') are a scale-less and elegantly elongated family of Southern Hemisphere freshwater fishes, with their greatest diversity in south-eastern Australia and New Zealand. The adults are mostly found in fresh waters though some that have a marine juvenile stage breed in estuaries, and the common galaxias (see chapter 17) is perhaps the most naturally widespread freshwater fish in the world.

Most galaxiids live in cool or high-altitude inland waters, with specialised species found nearly up to the summit of Mount Kosciuszko, while the flathead galaxias is widespread through the seasonally warm waters of the Murray River system. Many new fishes in this family have only recently been recognised, and even the barred galaxias with its distinctive burnt-orange colour and irregular black bars was regarded as just another colour form of mountain galaxias not all that long ago; there are many others that are just as intriguing.

In his recent revision of the mountain galaxiids Tarmo Raadik has found 15 new species, bringing the number of galaxiid species in south-eastern Australia to 35. We will probably never know how diverse this Gondwanaland family was before the introduction of trout, as these aggressive carnivores not only invade similar habitats and compete for food, but also feed on the galaxiids themselves. The galaxiid species that survive under this intense pressure usually remain most abundant in shallow areas with abundant shelter or, in the case of the Tasmanian *Paragalaxias* species, which are a touch more goby-like than the rest of the family, prefer to live and feed among rocks and rubble.

An unnamed species of mountain galaxias from south-western Victoria.

them underwater and at night. Like many galaxiids it spawns in fresh waters, with the young washing out to sea and returning mixed in with the spring whitebait migrations.

Frogs and reptiles

For all their natural associations with water, there aren't all that many frogs that live or breed in streams or their backwaters. The reason is obvious – such waters are likely to have fishes present, and though the dangers they pose to tadpoles have sometimes been exaggerated, it would only take a small school to finish off a whole season's breeding for a pair of frogs, just by eating one newly hatched tadpole each. Despite this some of the largest frogs are mainly found along streams in forested places, and the most striking of these are the barred frogs.

Named for their striped hind legs, these prominent-eyed and handsome creatures are camouflaged in muted browns and off-greens that make some of them look very leaf-like. Their tadpoles are strong swimmers even in moving waters, and are large enough that they probably aren't all that attractive as prey for most fishes. However, the mating frogs take additional precautions to minimise predation, either creating tiny pondlets of their own in the shallowest places where fishes are least likely to go, or by kicking their eggs out of the water so they stick just out of reach until they hatch and the active young can disperse.

As streams slow further a greater diversity of animals begins to appear. Around the higher reaches the only reptiles that are likely to be found near the water's edge are various small species of skink, but in the south-east a much larger lizard suddenly appears in substantial numbers as the waters step down to a meandering pace. The eastern water dragon spends much of its time in trees overhanging the water or foraging along the shoreline, but readily leaps into water and can remain submerged for long periods if disturbed. Left to its own devices it is unlikely to go swimming spontaneously, and although it may sunbathe part-in and part-out of the water it prefers to hide in burrows or dense cover on slightly higher ground.

Eastern water dragon.

13

Slow rivers in an ancient land

When rivers reach lower-lying ground they slow down and start to drop their sediment load gradually, though the finest particles may be so small they will never settle, so the water may remain tinted or even clay-coloured. The rivers that creep across the flat lands of inland Australia are the slowest of them all, winding thousands of kilometres through mallee and desert scrub, and forming natural corridors that link some very different places and types of wetland across a largely arid landscape.

Although many smaller river systems trail across the inland, the Murray–Darling is by far the most extensive, draining around a seventh of the entire continent. Its two main rivers originate in very different conditions, with the headwaters of the Darling sprawled across large areas of southern Queensland which may be drought-stricken at times, or carry great floods when monsoonal rains spill further south than usual. The waters of the Darling remain relatively warm even as they move southwards through desert areas, though in dry times when water levels are low the temperature may fluctuate more radically at the colder times of year.

Most of the Murray's tributaries start off in higher and much colder country where seasonal rains are a little more reliable than in the dry country to the west, and heavy frosts and snow are not uncommon. There is no great and dramatic whirlpool where the two now-sluggish rivers meet not far from the South Australian border, but the murky waters here mark an unusual ecological transition point.

Tall flatsedge growing along the Murray River.

The Darling has long been a highway for aquatic animals from warmer climates, giving them time to reach and dip their metaphorical toes into the cooler waters of the south, and various subtropical fishes, including gudgeons, a rainbowfish, a glassfish, a catfish and even a grunter have worked their ways upstream along the Murray from this river junction. Though these warmth-loving animals were well established and widespread along the Murray and its tributaries before weirs dropped water temperatures still further, most fishes from families that evolved in colder climates haven't managed to make the reverse journey up the Darling for any great distance, and nor has the Murray crayfish.

Even in its present debilitated state the Murray–Darling has a certain austere beauty, and I have often thought about the paradise it must have once have been, before the arrival of European settlers. In accounts from past decades and even in the memories of long-time residents the waters were reported to be much clearer than at the present time. Though this may seem doubtful to anyone who is familiar only with today's gritty-looking flows, the first Murray crayfish I saw at the tender age of eight was 3 metres deep in the Ovens River, too deep for me to dive down to without a mask and snorkel, yet its bright, white claws were crisply defined against the algae-covered rock it strolled across.

Plants: river red gum

The river red gum is the most widespread eucalypt in inland Australia, and also the dominant and most conspicuous tree along tens of thousands of kilometres of rivers and backwaters, including short-lived desert streams where the water may soak into sandy soils within days of heavy rain. Tolerating extremes of both flood and drought provided neither lasts too long, like many widespread wetland plants its primary adaptations are designed to spread it efficiently to other places that flood intermittently, rather than specialisations for growing in a particular climate or soil type.

Flooded river red gums surrounded by giant rush.

Though red gums are always found in intermittently flooded places, including backwaters and billabongs where they may be the only type of tree present, paradoxically they are not used directly by most types of wetland animal, though they provide both living quarters and corridors for the spread of many more terrestrial animals from possums to carpet snakes across landscapes that would otherwise be too dry for them. Red gums along the larger rivers reach the greatest size, developing hollows as they age and eventually falling to become underwater snags large enough even for a spawning pair of Murray cod. And these underwater hollows are long-lasting too; submerged red gum timber will remain in good condition for anything from decades to centuries, as long as it doesn't dry out too often.

Murray cod.

Fishes of the inland rivers

The sheer scale of the combined Murray–Darling River catchment has allowed the development of a relatively specialised fish fauna here, including the Murray rainbowfish and flathead galaxias, representatives of groups that usually live much further north and south respectively. In the northern reaches of the Darling River many fishes that have also adapted to cooler conditions remain much closer in appearance to their ancestors and close relatives in the streams of south-eastern Queensland. In the case of Murray rainbowfish northern populations still look markedly similar to their close relative (and probable ancestor), the crimson-spotted rainbowfish, and may be hybrid forms between two species that would never be confused at the extremes of their range.

Other fishes of the Murray–Darling are more uniform in appearance wherever they occur, including the Murray cod, which is Australia's largest freshwater fish. This giant remains moderately common in some places though it rarely now reaches the hundred kilogram and larger sizes recorded in the past. For a large-mouthed predator that can swallow its smaller siblings, possums and ducks whole, it is a curiously timid fish that mostly hides in the deeper pools of slow and muddy rivers.

It isn't hard to understand the long-term evolutionary logic that has created this behaviour. Most smaller fishes must breed prolifically within their first year of life because they have only a small window of opportunity, perhaps just a few weeks during which conditions are warm enough, or there is enough tiny food available for their fry. The males of such species, including the many rainbow-fishes, are a good example, often displaying ostentatiously even when predators are around, because by the time the next ideal breeding season rolls around they may already be too old to attract mates.

As with desert fishes that migrate upstream during floods, into places where most of them are likely to die as the waters dry out, it is the behaviour of the successful, risk-taking males that shapes the nature of the smaller species as a group. By contrast, a big female cod can produce 40 000 eggs each summer for decades, but she won't be able to *start* spawning for many years. This impressive number also tells us that though she has the potential to produce a million or more offspring over her lifetime, only a tiny proportion of them are likely to reach breeding size themselves, and it is likely to be the most cautious cod that live long enough to produce enough offspring to replace them in later years.

The closely related but much smaller trout cod had disappeared from nearly all of its comparably wide range not all that long ago, probably as a result of habitat destruction. A fish of clearer, faster-flowing tributaries with gravelly rather than muddy bottoms, it was less able to cope with the increased sedimentation caused by clearing, and the loss of shelter and breeding habitat as snags were removed to supposedly improve rivers. Apart from a few relatively subtle characters such as the overhanging upper jaw and relatively unpatterned head, trout cod can easily be confused with their much larger relative and it wasn't all that long ago that their existence as a separate species was doubted.

Genetic research is revealing an ever-increasing range of such cryptic fishes, and as new species turn up from what once seemed a reasonably well-mapped

Hybridigen males of a Murray–Darling carp gudgeon.

Barcoo grunters.

freshwater fish fauna, even the most conservative estimates of just how many new fishes remain to be named suggest their numbers may double in the foreseeable future. One of the most intriguing of these closely related yet surprisingly variable groups is the inland carp gudgeons of the Murray–Darling, once regarded as a single species with diverse local variations.

It soon became obvious that at least four apparent species were being lumped together under a single name, but the results of recent genetic studies are even more curious, as some of these groups include self-sustaining first-generation hybrids of true species that seem to be able to get by even if their parent species have disappeared from the same waters! These genetic parasites (including hybridogenic forms which are mostly male, and others which are mostly female) must mate with a closely related species to produce offspring, even though they may not have any actual use for many of its genes.

Larger fishes such as the golden perch, found through much of the Murray–Darling and also further north in arid subtropical regions, may also turn out to be more than one species, with a northern form able to tolerate even greater extremes of flood and drought than those in the larger and more reliable rivers to the south. It shares these slow-moving rivers and turbid waterholes with Barcoo grunters, with a reputation as a leathery and poor-quality edible species when caught from the wild. Yet more recently this fish has become one of the minor princesses of the aquaculture industry, as one of its adaptations to these unpredictable waters is the ability to store fat to make up as much as 15% of its bodyweight.

Turtles and invertebrates ... but few frogs

With several species of large, predatory fishes widespread through the Murray–Darling, it isn't surprising that frogs are never found in the broader and more open reaches of water, though they may be abundant at times in well-vegetated backwaters and more isolated swamps not far from the main rivers. Large fish may still be found in some of the deeper billabongs at times as their young may be washed in during floods, and the three species of freshwater turtle found through the catchment can stroll between the main rivers and nearby swamps at will, but the backwaters are far more likely to offer some protection to both frogs and their tadpoles.

Turtles were probably much more abundant through all these waters before the introduction of foxes which dig up many of their nests, but they can still sometimes be seen in small numbers lined up along logs just above the water, sunbathing to clean up fungi and other pathogens that could damage their shells if they remain wet for too long. The short-necked Macquarie turtles are most likely to indulge in this habit for long periods, with common snakeneck turtles likely to return to the water more often, so the algae that covers their shells remains lush and green.

Primitive though they may seem, turtles have a keen awareness of their surroundings, and it is difficult to sneak close enough to get a photo of an entire group out of water, because one of them is sure to see you and dive with the others just a fraction of a second behind. Broad-shelled turtles are by far the largest of the

Murray crayfish.

Southern freshwater prawn.

long-necked species, curious and relatively small-headed beasts that can reach almost a metre in length with their necks stretched out. The neck is even more prominent in the young of this species, which add to their novel appearance by clustering in loose schools at times.

The Murray crayfish is the second-largest freshwater crayfish in the world, and is widespread through the Murray itself up to some very high and cold altitudes, but shows little sign of wanting to move upstream into the warmer Darling River. This is a huge and handsome beast with white claws, and neat but formidable arrays of upright body spines; larger animals are best handled with gloves, as their effect on an unaccustomed bare hand can be a little like grabbing the wrong end of a power tool. Of all the native crayfishes only the common yabby, also found through these waters, is more widespread, but the smooth crayfishes are most abundant in backwaters (see the next chapter) because they just don't have the spiny armature that helps protect the Murray crayfish from larger predatory fishes.

Glass shrimps are also widespread and may even be present in the main rivers, but the common shrimp-like animal here is the southern freshwater prawn, with its arms as long as its body. Unlike the closely related cherabin of northern waters these prawns are too slender to be easily confused with crayfishes, and they are also often much more abundant. Periods of prolonged flooding seem to simplify the underwater landscape of the rivers, laying down a carpet of silt that many other animals avoid, but with potential competitors and predators reduced in numbers the prawns breed up and pick their way across the soft sediments in great numbers.

Freshwater mussels from the Murray River.

And shallowly buried in the silt and sand beds are freshwater mussels, with only their siphons showing most of the time as they filter water through their amorphous-looking digestive systems. The species in these rivers grow quite large and were probably an important and easily harvested food source for Aboriginal peoples, as they are easily found by digging along the surface sediments with your toes. Their flavour is too bland for most people today unless they have been soaked in spices before cooking, but these high protein snacks still feature regularly in water rat middens.

Waterbirds: drifters and stayers

Unlike fishes and crustaceans, the waterbirds usually found around the Murray–Darling haven't evolved into any distinctive regional species, probably because these have always been highly seasonal waters and flight is the easiest option for dealing with periods of drought. The behaviour of the Australian shelduck is a good example, as it is often found scattered through many parts of the catchment of the Murray River while there is enough water around. However, before the major moult when this species becomes unable to fly for some time, most shelducks along the northern side of the Murray move further south to where there are likely to be more permanent waters, and enough food to carry them through.

The huge extent of well-watered wetlands during good years, when heavy rains fall in the north or run from the mountains to the east, means that the basin as a whole can abruptly become a gigantic nursery for many common species, even

Magpie geese.

after years of drought. Even chronically restless birds such as whistling ducks may move well south of their usual range during these times of abundance, though they always return to the north of the continent to breed around the beginning of the monsoon season.

The magpie goose was also widespread in the south, including along parts of the Murray and its tributaries, until it was eradicated by hunting pressure, poisoning (the geese love grain) and loss of many of its breeding grounds to drainage. These were usually shallow, well-vegetated wetlands with a reliable and abundant supply of grasses and seeds, yet even here the geese would have to abandon their young if the water and food disappeared too quickly, leaving little chance of raising them successfully. Their high degree of adaptation to the ever-present threat of drought has also affected the way they moult, so unlike shelducks magpie geese shed only a few flight feathers at a time and never completely lose the ability to fly.

Hardhead ducks are still widespread across most of the continent though their main strongholds are in the Murray–Darling region, and they have become much less common in many of their more eastern haunts. Yet this far-ranging species has also appeared at times in substantial numbers in Indonesia and New Guinea, and

Azure kingfisher.

was fairly common and breeding in New Zealand for half a century before disappearing from there. Their preferred habitats are said to be deep lakes as they are diving birds that collect some of their food from the deeper waters, though they are mostly herbivorous and often build their nests among cumbungi and tangled lignum, both plants of shallow and often very seasonal marshes.

Regardless of conditions across the whole catchment in any one year, there are some birds that can be reliably found in small numbers wherever there is reasonably permanent and moderately clear water, of which the azure kingfisher is perhaps my favourite. Often hard to see from any distance, its distinctive high-pitched whistle is easily recognised once you have heard it, and following the call you are likely to find the electric-blue bird on one of its favoured perches. A tiny bird with a big personality, it will sit virtually motionless with only occasional shifts of its head as it watches intently for any movement in the water below, before suddenly lifting off and whirring to a new vantage point.

Human impacts: why the bunyip disappeared

Though still attractive in a wet year, much of the Murray–Darling is now just a series of regulated, degraded waterways, and wetlands that turn into dustbowls though the lawns of nearby country towns are still being watered from the various tributaries. By the end of the prolonged drought that opened the 21st century, waterbird abundance through the Murray–Darling had dropped an estimated 80% from the early 1980s, and 75% of the wetlands in Victoria that would normally be filled by the rivers were dry. Yet despite increasing concerns over water quality and habitat, some industrial-scale farms were still allowed to take more irrigation water from these rivers than has been allocated for environmental flows for the whole system.

Weirs that regulate the flow of the Murray River were originally intended to allow paddlewheel steamboats to operate even through droughts.

Regulations governing water use in the Murray–Darling are a mess that has been caused partly by the conflicting interests of the states the major rivers pass through, with South Australia getting whatever poor-quality water is left after everyone else has already dipped deep. Weirs have had multiple effects on habitat, including making it possible to divert much of the natural flow into surrounding lands for irrigation, which will eventually destroy much of the farmland that depends upon it as salinity increases, as has happened in the historic past elsewhere. Even the abundance of many native fishes has been reduced in part by weirs, with the release of colder waters from their depths discouraging breeding.

The least familiar casualty of the advent of weirs is the bunyip, as most reports of this animal in the past clearly refer to seals. These would have been an astonishing sight to Aboriginal peoples who lived many hundreds of kilometres from the sea, and even as late as 1850 a large group of seals was seen swimming upstream along the Murrumbidgee at Gundagai, 1200 kilometres from the mouth of the Murray. Ignoring spurious bunyip reports such as the deformed skulls of colts and calves, an Aboriginal sketch of an emu drawn by some humourist making fun of a gullible whitefellah, and the ridiculous suggestion that Aborigines couldn't identify the booming calls of bittern so they made up a myth to explain them, seals tidily fit the bill for most bunyip sightings. This even includes its abrupt disappearance once the first weirs were built, though to be fair the depredations of sealers had already reduced seal populations in Bass Strait dramatically, making it much less likely that strays would find their way inland.

Irrigation canals draw off a considerable volume of water from the Murray–Darling, increasing salinity through evaporation from their open waters.

14

Floodplains, billabongs and backwaters

Rivers and streams not only link many types of wetlands, but they also create new ones along their path, particularly during floods. Though the main channel may run fast when the waters are high, the spill to either side slows down as it spreads, shedding sediments to form broad and fairly flat areas with deep and often nutrient-rich soils, known as floodplains. These may be relatively narrow in a valley or spread many kilometres beyond the main course of the river, though sometimes there is no main channel, and the moving waters break up to form many shallow streamlets that braid in and out of each other.

Rivers tend to wander across their floodplains, gradually carving out great loops as the faster-moving waters on the outside of the curve cut into the shoreline, and sand or silt build up along the inside. During floods the river will try to take a more direct route across its floodplain, and eventually it may cut a new channel, leaving a loop isolated from the main flow to form a backwater. At first this will still be linked to the main channel so river animals can come and go, but with little water movement to keep it open the entrances gradually silt up and it becomes a curved lagoon known as a billabong in Australia (or an oxbow lake overseas).

Inland billabongs are usually so surrounded by trees that it isn't easy to see how they formed in the first place, and ironically it is much easier to follow the process when looking at tidal sand in a broad estuary (see chapter 17), because there is no

Backwaters are a favoured habitat for freshwater turtles such as this common snakeneck.

vegetation to interfere with the view. On the immense floodplains of the Top End things may be even less clear, with no obvious boundaries between streams, wetlands and backwaters during the wet season.

Even if the now-isolated billabong dries out from time to time, it will still be refilled and restocked with aquatic animals from the river during floods, and at other times will also attract a different range of animals that aren't keen on moving water. As sedges and reeds establish and submerged plants appear in clearer waters, a new and often complex ecosystem develops with its elements drawn from many places, often becoming strikingly different from the river still flowing not that far away. In time the billabong ages as sediments continue to wash in, turning into a marsh and ultimately into a strip of seasonally boggy ground.

In some parts of Australia larger rivers are now so regulated that there is little chance that new billabongs will form in the future, and as the older ones fill up and dry out a whole range of slowly changing habitat types and some of the unique animals that favour the rivers' different ecological stages may disappear. The Murray River in particular has its flows so tightly regulated that there is little chance it could cut a new channel again; landowners adjoining the river are not only forewarned weeks ahead of water level rises, but also precisely how few centimetres it will be allowed to rise.

An ageing backwater near the Murray River, now reduced to a shallow claypan with dense aquatic growth and fringed by inland paperbarks.

Human impacts: weeds

Weeds are an ever-increasing economic and environmental problem that already costs billions of dollars annually within Australia alone, yet new species continue to appear in new locations every year. Although floating plants such as water hyacinth are among the worst of these, and a few species can grow thickly underwater, their greatest impacts are usually in shallower waters such as sluggish streams, backwaters and marshes. Relatively few introduced weeds grow in undisturbed wetlands, but those that do are real problem plants that smother indigenous species, and are useless as habitat for anything but mosquitoes.

Many of these have been introduced and spread through the nursery and aquarium trades, others as supposedly useful plants for agriculture or water treatment. Olive hymenachne was deliberately released and spread as a pasture grass by a Queensland government agency as recently as 1988, despite warnings of its disastrous environmental potential. Within a decade it had become a major environmental weed in Kakadu where it can choke out all native plants in shallower streams and backwaters, and current efforts to curb this weed in Queensland are restricted to just keeping it from spreading further, because there is little chance that it can be eradicated now.

But even indigenous wetland plants can potentially become serious weeds, particularly if introduced outside their native range, or if they are in nutrient-rich situations where their growth becomes explosive. In the image below a smothering red carpet of the native Pacific azolla has reduced oxygen supply to a shallow pool already choked with introduced weeds.

A narrow backwater choked mainly with introduced weeds, including the aggressively invasive Mexican waterlily, cabomba and cat-tail.

Insects and crustaceans

The dense stands of emergent sedges and grasses around many backwaters make ideal habitats for various dragonflies and damselflies, and the jungle of submerged plants in the shallows is often filled with their larvae. Around tropical backwaters dragonflies of many colours and patterns regularly form chaotic flights that dodge and spar ceaselessly, casually confident that they can outfly any predatory bird. Even in the south there are years when many dragonfly species suddenly become abundant for no apparent reason, and one season we had around 400 dragonflies of several species fighting and mating around our swimming dam, over a 2-month period.

As the males are territorial, suitable places where they can wait for a passing female become scarce under such conditions, and any new territory that becomes available is highly sought after. I was amused when one particular male seemingly laid claim to a new pond I was digging before there was any water in it, flying around inside the excavation proprietorially whenever I took a break, and moving in permanently the day it was filled and a few pots of plants had been placed in it. Rather than spoil his romantic prospects I didn't introduce any fishes, and sure enough within 3 weeks there were already tiny mudeyes creeping along the stems of submerged plants.

If large mudeyes are present in any numbers smaller fishes become scarce, because those which take shelter among plants are often eaten, whereas any that stay further out become relatively easy prey for waterbirds and larger fishes. But if the fishes are already abundant the balance can go the other way, as even the smallest species will greedily feed on mudeyes and nymphs as soon as these hatch, and while the nymphs are still tiny small fishes are in no danger as they scour through beds of plants in search of these morsels.

Backwaters are a paradise for many other insects as well, though most of these are nowhere near as bold as the dragonflies and you will need a net to find them. And among the mass of hopping, skipping and sometimes biting insects, spiders with impressive fangs and attitudes, not to mention some very active leeches each netful will bring to the surface, you are also likely to find various small crustaceans. If the backwater dries out regularly these will mostly be copepods, water fleas and the other groups already familiar from earlier chapters, but shrimps are unlikely to be present and the only larger crustaceans will be the young of smooth crayfishes.

These are all lowland and floodplain animals that tolerate a wide range of temperatures and salinities, usually retreating to deep burrows during periods of

Spanner-claw yabby.

drought. The common yabby of the east is the best-known species and is also the most widespread of all our native crayfishes, now even more so due to deliberate releases elsewhere as well as escapes from aquaculture. Some of its larger relatives are also aquaculture animals as they are faster growing and more tolerant of warm waters than most other crayfishes. Marron and redclaw are basically free-ranging animals of permanent waters, and although yabbies burrow readily they aren't as drought tolerant as still others of their relatives, which prefer even more ephemeral wetlands.

The most intriguing of these is the spanner-claw yabby, a secretive species which seems to spend much of its life in deep burrows from which it is very difficult to extract. Now mostly found around the general vicinity of the largest river red gum forest left in Australia along the Murray River, it may have once ranged more widely. Although regarded as fairly rare this may just reflect the difficulties of studying a crayfish that is easily caught only after floods, when it becomes difficult to make your way into its isolated habitats.

Nearly all spanner-claws collected have been males, with their deep, broad claws that look (and possibly may work) much like a spanner, but for all their fierce appearance they aren't notably aggressive if kept in an aquarium as long as they are kept reasonably well fed. This handsome crayfish hasn't even been formally named at the present time, and almost nothing is known about its biology, yet the powerful claws may hint at a highly specialised lifestyle unlike that of any other crayfish known.

Some other smooth crayfishes also have soft mats of what looks like fur on the inside of each claw, but in the spanner-claws this is exceptionally thick and dense, and combined with the shape and close fit of the fingers of the claw creates a strong impression of the curved jaws of a baleen whale. In the case of the whales there is no doubt what these close-woven structures are for: they are specifically designed to collect minute, planktonic animals from any water that passes through them.

It is possible that these curious-looking claws work in a similar way, and my suspicion is that the fur is used to comb minute animals from both living and dead vegetation in the ephemeral waters where they live, perhaps even from the water itself if there are enough minute animals swimming about. The ability to feed on the tiny planktonic animals that swarm in most recently refilled wetlands, along with being able to survive underground for long periods whenever those wetlands are dry, should be a winning combination, though it begs the question of why spanner-claw yabbies aren't more common or widespread.

Fishes, frogs and reptiles

Backwaters that dry out seasonally make a secure breeding place for animals that can reach them via floodwaters, but are a risky environment for most fishes. Among indigenous fishes the salamanderfish of south-western Australia

Eastern little galaxias.

(see chapter 7) is by far the best adapted to drought, and may make its way between isolated pools in the wide and peaty heathlands where it lives during floods, while water is running in braids that link many parts of these low lands.

It is found in many of these places with two tiny galaxiids that may share some of its ability to survive hot and dry conditions, though these are more often associated with streams and there doesn't seem to be any evidence that they can survive the complete drying out of any pond they may be trapped in. In the south-east their close relative, the eastern little galaxias (which may actually be two species, found on the eastern and western sides of Port Phillip Bay), is also most often found in streams that may spread more widely during winter rains.

My observations on this fish over two decades suggest that it remains most abundant where it can breed in shallow, winter floodwaters, where predators such as mudeyes, which can wipe out every young galaxias in a pond that remains full at all times, are rarely found. Unlike its western relatives this fish does seem to have some strategies that can prolong its survival in more ephemeral waters, including retreating down the burrows of marsh yabbies, or lying low in wet places under logs and other insulators that protect it from the worst heat of the sun.

Rainbowfishes have no means of surviving in drying ponds if they are trapped, but there are always plenty of survivors in the deeper pools of streams and in the backwaters associated with river systems, forming schools wherever there is a steady flow of water. Restless young fishes readily move upstream and are even able to jump low obstacles, but, as has been seen in earlier chapters, these drying waters often become a bonanza for predators, with the doomed rainbowfishes becoming an essential ingredient in the survival of other species.

Away from the larger predatory fishes in the main river channels and with dense thickets of plants to shelter tadpoles, frogs begin to appear, though they are usually heard rather than seen. It can be frustratingly difficult to locate a calling male at close range because the call can't necessarily be pinpointed without a second person to help you triangulate on its source, and also because the frog is likely to shut up as soon as it realises you are nearby. There are exceptions: many treefrogs take to captivity readily, and of all of them Dahl's treefrog in the Northern Territory seems chronically inquisitive, peeping out from under waterlily leaves minutes after being disturbed if you remain still.

Slow rivers, clear backwaters and billabongs are the preferred habitats of freshwater turtles; such waters offer a wide array of foods to suit all sizes and ages, from insects to smaller fishes which they stalk before snapping up with a last-minute lunge. Their thick-shelled eggs are well designed to prevent them from drying out, but they will drown if flooded so female turtles go to a lot of trouble to bury them well above any likely flood level. This may involve walking hundreds of metres and climbing up fairly steep banks, an inconvenience for the mother but even more so for the hatchlings once they have dug their way out of the nest.

File snakes are perhaps the most specialised aquatics among native freshwater reptiles. Named for their rough skins, these live-bearing, nocturnal creatures are graceful, sinuous swimmers in water, but turn into flaccid, almost inanimate tubes

An Arafura file snake tastes the water.

if taken out of it. Feeding primarily on fishes, they have evolved for the great floodplains of the far north, following their prey into larger billabongs and backwaters, but avoiding smaller pools where they would be almost as surely trapped as their prey. Non-venomous and reasonably easily caught, they are still common despite being a favoured food for Aboriginal peoples, an indication not so much of their fishing skills but more of the abundance of the small fishes they feed upon.

Rainbowfishes: flamboyant diversity in the north

Rainbowfishes are the most abundant and widespread family of smaller freshwater fishes in Australia, preyed upon on a grand scale by diverse animals from larger insect larvae to other fishes, many waterbirds and turtles; they are even the single-most popular fish family in the diet of the freshwater crocodile. Although there are more species in New Guinea, the family is widespread through tropical northern Australia well south into desert regions, with one species found as far south as central Victoria. Several species or variants are shared with New Guinea, including the threadfin rainbowfish, a tiny creature with extremely elongated fins, though in Australia it is restricted to the tip of Cape York and a corner of Arnhemland.

To offset their high casualty rate in the wild, rainbowfishes are prolific breeders in warmer conditions, and their combination of vivid colours, striking patterns and the lightning-fast sparring displays between males has made these one of the most

Few wild rainbowfishes reach the size of this old chequered rainbowfish.

sought-after families of freshwater aquarium fishes worldwide. The large and exceptionally deep-bodied males often displayed in public aquariums rarely live to reach such a size in the wild, and their colours have often been enhanced by selective breeding over many generations, but some of the fishes netted from creeks and backwaters will already be as gaudy as any coral reef fishes as soon as they have settled down to aquarium life.

Male rainbowfishes often joust, but sometimes the sparring turns into serious blows as in these crimson-spotted rainbowfish.

Waterbirds

The wealth of plant and animal life in a healthy, seasonal billabong or backwater can be astounding, and for nearly every living thing present there will almost certainly be a waterbird that feeds on it. Some are so highly adapted to conditions in these still waters that they are unlikely to survive elsewhere, particularly the comb-crested jacana of northern Australia (also known as the lotusbird). Tripping carefully on its extremely long toes that allow it to walk across waterlily and lotus leaves, it turns these up to feed on various invertebrates attached to the underside, and also feeds on the seeds of the aquatic plants it is always found among. Even its floating nests are made from aquatic plants, and the black-scribbled patterning of its eggs is cryptic against blackening, dying sedge stalks.

Lotusbird and Dahl's treefrog habitat: a waterlily-filled billabong in the Top End.

Rails, crakes and bitterns are far more widespread but are much less likely to be seen, lurking in reed beds and only emerging briefly, usually inconveniently close to dusk if you have been waiting all day behind a camera. The diet of rails and crakes is a combination of invertebrates and vegetable matter including seeds, just the sort of stuff likely to found in the shallows of a billabong, but they are presumably fairly flexible in their diet as they can be found across a range of wetland types, including sometimes saltmarshes. Apart from hiding, crakes will also dive to escape attention, and even their nests, cryptic messes of vegetation barely raised above the water's surface and usually overhung by sedges, are very hard to find.

Ibises are better known for their communal breeding habits, sometimes forming huge colonies in paperbarks when these have been flooded. However, they will also nest in much smaller groups with a dozen or so pairs making untidy nests of flattened spikerushes or other sedges in more permanent backwaters, presumably relying on the deeper waters these plants prefer for protection as the nests may not be far from shore. Any increase in the risk of predators finding the young must be partly offset by the convenience of being able to pick your food from the seething hordes of aquatic insects and other small animals massed in the surrounding waters.

In every billabong and backwater there are invariably ducks, including all the generalist species that will make themselves as readily at home in lakes as in ponds. For herbivores, omnivores or hunters, there is likely to be a good supply of food

Buff-banded rail.

here through the critical weeks of the breeding season, around spring in the south or as wetlands further north begin to dry out after the wet season. Yet some ducks have special needs of their own, and prefer even shallower and more seasonal marshes and swamps further from the maddening crowds, as will be seen in the next chapter.

A mixed flock of ibises at the edge of a drying billabong.

15

Marshes and swamps

It isn't easy to define or categorise the many types of shallow, boggy and often ephemeral wetlands that are loosely tied together as marshes or swamps, yet these can be among the most productive and varied of inland aquatic ecosystems in terms of the many species to be found in them. Often so nutrient-rich that they may be mostly hidden under a dense blanket of plants, marshes look like solid ground unless you make the mistake of trying to walk across their pitted and unpredictable surfaces. These are among the most difficult places to study the many types of animals which swim through their overgrown waters, burrow in their soils, or tunnel into the leaves and stems of plants.

Some marshes are filled by rain that goes nowhere and evaporates within a few months, so they become slightly saline over time. At another extreme others may be kept full throughout the year by pure fresh water flowing not far below the surface, as in Eighteen Mile Swamp on Stradbroke Island. This is effectively a stream flowing steadily beneath the high dunes that make up most of the island, surfacing only when it reaches the lowest ground, where its clean and clear waters are pumped to the nearby mainland as part of the suburban water supply there.

Swamp plants: diversity and interactions

Plants shape the ecology of most marshes and swamps, in turn selecting the animals that are likely to be attracted to them, and their diversity is obvious even at the most superficial glance. From carpets of low-growing grasses, sedges and herbs that may have a chance to flower or set seed only during the wettest years, to tall shrubs and palms that grow in places that are reliably flooded for part of the year,

Swamp fern scrambling up the trunks of flooded gebang palms in the Top End, an ideal habitat for treefrogs.

they create three-dimensional arrays of emergent habitats that can sometimes tower well above ground, attracting more terrestrial animals from pythons to birds, few of which need to venture near the waters far below.

However lushly the plants in a marsh or swamp may grow there are often only a few dominant species present, and these may be combined in surprisingly different ways even in wetlands just a few kilometres apart. In many of these cases the first plants to arrive have probably found ways of keeping later arrivals at bay, producing a variety of allelopathic (also known as antagonistic) chemicals that range from simple organic acids, alcohols, tannins, alkaloids and sulfides to long-chain fatty acids, which are released into soil or water as deterrents to potential competitors. Allelopathic chemicals are also used against animals that attempt to feed on a plant, but as they are usually manufactured only in response to an attack of any kind, it may take some time before the plant can build up enough of any defensive substance to have the desired effect.

Plants: sedges, reeds and rushes

Grass-like plants, including sedges, cumbungi, grasses and the various families known as rushes, are the dominant plants in many kinds of wetlands, from the fringing vegetation of lakes to the dense and tangled thickets that may completely cover marshes and swamps. Sedges are closely related to grasses, and are usually the most diverse and abundant grass-like plants in wet places, though several groups of true grasses may also cover extensive areas in shallow marshes. The common names of many of these plants aren't a useful guide to the families they belong in; for example, spikerushes, clubrushes, twigrushes and leafrushes are actually all sedges. Most of the true rushes and cordrushes are fringe dwellers, growing near the water's edge and tolerating prolonged floods, but they are usually most prolific in places that dry out for some of the year.

Common reed is the most widespread of aquatic grasses, found worldwide in dense, single-species stands that don't provide much in the way of useful habitat for

Sedges and water-ribbons surround a deeper pool in Eighteen Mile Swamp.

most larger animals, though fishes will forage through them at high tide, and reed-warblers and bitterns nest and feed around their fringes. The clamorous reed-warbler is found throughout Australia and weaves its nests among reeds (and also cumbungi) stems, where they are protected from most predators by the water below. Reeds can cover such extensive areas because they are tall enough to shade out many potentially competing plants, and also tolerate a greater range of conditions than most wetland plants – from drought to the upper tidal reaches of estuaries.

Dusky moorhen nest constructed from tall spikerush woven through common reed.

Invertebrate lifestyles

Plant diversity in a marsh or swamp also generates animal diversity, and there are probably very few species of plants in these places that aren't nibbled on at times by at least one or two types of invertebrates. Many water snails will feed on plants, but as poor travellers and with a preference for reasonably permanent waters there may only be one or two species of these present in a swamp, and these are just as likely to be scavengers, algae eaters or detritivores as plant eaters.

Most of the plant-eating invertebrates are likely to be insects, often the larvae of terrestrial species that stay out of reach of the more predatory aquatic species below. The varied sword-grass brown is one of the most conspicuous insects in the south-east: a big and bright butterfly that loops erratically not far above the ground as the males patrol the sawsedges their young will feed on, always alert for newly emerged females. For all of the considerable variation in colours between populations there is also a surprising degree of consistency, and the dark colour-morph which is common in my local area looks no different from some I have seen a thousand kilometres away.

Few of the insects in the waters below feed directly on plants, and though many of them feed on the organic matter that builds up rapidly, or on bacteria and other minute organisms within the muck, many of them are carnivores. Backswimmers and beetle larvae are often found among the plants, some of them so finely sculpted they would make attractive jewellery if used as a mould, and there will

Dark morph of the sword-grass brown on its host sawsedge.

A random sample from a shallow pool in a marsh with numerous species of insects or their larvae, and the ghostly shadows of copepods and water fleas.

always be at least one species of caddisfly larva present, though these are superbly camouflaged in the shells, husks and plant-offcuts they choose to hide in, according to the preferences of their species.

Unlike wetlands on floodplains, where silt-bearing floods gradually fill billabongs in over the years and smooth out the landscape, marshes and swamps are more tenuously connected to flowing waters so their underwater landscape may be much more variable. The deeper pools may not dry out completely, allowing larger carnivorous insects and larvae to survive through drier periods, while the shallower areas may spawn great numbers of smaller crustaceans just as in other types of wetland; these multiply rapidly while they remain relatively isolated from the predators in nearby pools.

As it refills, a swamp becomes a very different place where both predators and prey can commute through connecting shallows to new areas, though larger insect larvae such as mudeyes are likely to stay in deeper waters unless food is scarce. Smaller predators, including damselfly nymphs and backswimmers, are less reluctant to explore further afield in pursuit of greater concentrations of the water

Red-and-blue damselfly.

fleas and copepods that are among their favoured foods, and will find their way to new pastures even through just a few centimetres of water.

The slender red worms that remain hidden in the sediments during the drier periods also emerge, retreating underground at the slightest disturbance, and the organic debris that has accumulated between pools becomes accessible to beetle larvae and waterboatmen, along with caddisfly larvae that may harvest plant fragments as both shelter and food. And when drier conditions return flying insects abandon the shallower pools, so any larvae trapped in the receding waters must transform into their winged forms whether they have reached their full potential size or not. Damselflies often become abundant around swamps and marshes around this time, but most species of dragonflies are likely to move on unless there are enough deeper pools to support their larvae until the rains return.

The nature of water: surface tension

Where the water's surface meets the air above, water molecules are more strongly attracted to each other than to the air so they form a thin, elastic membrane. Although barely noticeable to larger animals, this tougher 'skin' can become a barrier to many smaller life forms, or a platform for others. Underwater animals such as insects that must surface to breathe have to break through it to reach the air, which they do in several ways. Some poke a snorkel through to the air, with an opening often surrounded by water-repellent substances such as wax or grease, or oily hairs at the tip of a rat-tailed maggot's snorkel.

Insects and related life forms that live *on* the water's surface, from tiny, waxy springtails (see chapter 7) to needle-like water measurers that tread deliberately along the fringes, also use this approach to keep them afloat. On more open waters

Mating aggregation of water striders.

water striders are common, scattering to hunt by day but regrouping near cover at night, and sometimes forming large mating aggregations with the smaller males each clinging to a female. These spectacular skaters not only live and hunt on the surface, but they also communicate complex messages to each other by rippling the surface.

Whirligig beetles drift and spiral lazily near overhanging plants, speeding up into frenzied spinning and gyrating swirls if some small insect falls near them. Their feeding habits are amusing if not always attractive, as they bite out mouthfuls of their prey while simultaneously buffeting and shoving each other, or jumping on each other's bodies, all at high speed. Their perception of the world must be unusual; each eye is divided into two bulging halves, one half keeping watch underwater while the other scans above, and with the refracting effects of the surface tension layer they give the impression of having four separate eyes when seen from the front. Whirligigs find their prey and avoid obstacles through vibrations which they pick up through antennae in contact with the water's surface.

The separation of the divided eyes of these whirligigs is exaggerated by distortion through the water's surface.

Frogs, tadpoles and waterbirds

Marshes and swamps are an unfortunate combination for fishes – too far away from deeper and more permanent waters where most of the fishes live for easy access, and also prone to dry out too regularly for fishes to survive, even if they *could* make their way there. Here we enter into the realm of frogs, places where they can live and breed in comparative safety from many of their larger predators. It isn't entirely a paradise though, as there are usually plenty of insect larvae that are happy to feed on smaller tadpoles at least, but the plant cover offers reasonable protection from the greater appetites of waterbirds.

Many of the smaller frogs found in these places are cryptic species, with the common sense to remain out of sight during the daylight hours while their predators are most active. In the south-east the main exceptions are two closely related treefrogs that are among the very few amphibians that can be seen out in sunlight during the day. Growling grass frogs are found in the southern parts of this range and are easily recognised by their warty skin, with the smooth-skinned green-and-gold grass frog taking over the more northern range from eastern Victoria.

Although they were still abundant just a couple of decades ago, these large and distinctive animals have now disappeared from many of the swamps and marshlands where they once lived, probably victims of the ever-spreading chytrid fungus

Growling grass frog.

plague. Their absence has changed the very character of these wetlands for anyone who remembers them perched on almost every tussock near the water, or has heard their hoarse, grunting choruses from a distance. In the absence of these active hunters, which used to track down the smaller frogs that were a significant part of their prey by homing in on their calls, other more disease-resistant frogs have probably increased in numbers, with unknown flow-on effects for the future ecology of these places.

However abundant and desirable tadpoles may be as a food, they aren't easily caught among the dense overgrowth of many swamplands. Herons and egrets avoid wading through tangles of vegetation, unless tadpoles are so abundant that slow and laborious stalking becomes worthwhile, with each step needing to be considered carefully to avoid tangling their legs or alerting potential prey. Cormorants and darters are agile underwater hunters but are clumsy fliers, needing open spaces for take-off. Small pools don't offer much scope for them to manoeuvre, and even the most stupid tadpole will quickly retreat into the surrounding vegetation if a hunting cormorant turns its pool into the open-air equivalent of a washing machine.

Many ducks are better at take-off from enclosed waters though they give a strong impression of having to work hard at it and, although most of these are omnivores that will gladly snack on tadpoles when they are available, they are not exactly swift and efficient hunters. Freckled ducks – well-camouflaged birds whose

Freckled duck at rest.

white flecking on a dark background breaks up their outlines very effectively while they rest through the day – are perhaps the best adapted of them all for life in the enclosed spaces of swamps and marshes.

Emerging at dusk to feed, they are difficult to see in the lignum swamps and other heavily vegetated waters that are their preferred nesting sites except on moonlit nights, and though they must retreat to deeper and more permanent marshes during periods of drought it is unlikely that they thrive under these conditions. Given that their preferred diet is mostly made up of the tiny crustaceans and young insects that appear in their greatest numbers when shallow wetlands refill after a dry period, along with seeds of annual grasses and similar plants that crop most generously as swamps dry out, it is not surprising that this species breeds most successfully after periods of heavy rain.

For all of the difficulties of foraging among dense stands of plants, some birds have adapted spectacularly well to the prevailing conditions, and the purple swamphen is undoubtedly the best known and most widespread of these. With its large, long-toed feet it may look awkward when it strolls across your lawn, yet a swamphen can also climb with reasonable agility, and may use its toes to good effect in pinning down morsels grubbed out from among plants, or lifting them to

Purple swamphen showing off its handsome feet.

its beak. Feeding mainly on plant matter ranging from tender shoots to young tubers, it will also seize on any smaller animal too slow to escape when disturbed.

A swamphen's flight may seem clumsy, but it is well designed for near-vertical lift-off so that it can't be trapped too easily among the sedges and water-ribbons where it nests and feeds, yet it can also fly impressive distances without apparently needing to rest on land. With its glossy purple plumage and handsome red face shield, frequent flipping of its tail to show off the white underpants and its ear-splitting screeches, the purple swamphen must surely be the most characteristic and immediately recognisable of all marshland birds, by day or by night.

16

Freshwater lakes, lagoons and other dark waters

As the driest liveable continent Australia doesn't have many lakes compared to most other parts of the world, and many of these are markedly saline, as was seen in an earlier chapter. The freshest lakes are mainly found in relatively high country, or at the other extreme perched among silicon sand dunes, topped up by reliable rainfall that hasn't had the chance to flow through mineralised ground for too long. Their origins vary considerably from subsidence in already low ground, to damming by volcanic activity or carved out by glaciers in the higher country, and small lakes may even form behind landslips.

Most freshwater lakes in Australia aren't particularly old and their age can usually be measured in just tens of thousands of years, though some with more saline waters have been around (on and off) for millions of years. Freshwater lakes are almost invariably much more recent, which means there hasn't been time for a specialised lake fauna to develop except in a few favoured places. And the deeper a lake, the less varied the fauna found on its bottom sediments is likely to be, with a corresponding reduction in abundance.

Sediments in deeper lakes are less likely to be disturbed by currents or waves, so their waters are often poor in the nutrients needed to support anything much in the way of plankton. In lakes where there isn't much water movement oxygen levels can also be low, especially during the warmer months when a layer of warmer water sits at the surface, and this thermocline, where water temperatures suddenly drop dramatically as you go deeper, reduces gas diffusion into the deeper areas still further.

The nature of water: black waters – peat and tannins

The layer of peat in this subtropical sand-dune lake is all that keeps the water from draining away.

It may seem strange to link freshwater lakes with lowland rivers and coastal lagoons, but these include some of the freshest still and slow-moving waters in Australia, and a further unifying theme between these two apparent extremes is their often dark colour, which may look like over-stewed tea. This comes from tannins leaching out of peaty organic matter and, closer to the coast in places where it builds up behind sand dunes, peat creates new wetlands by sealing sandy soils that otherwise would never hold water.

Peat is basically partly decomposed plant matter formed in waterlogged or flooded soils with low oxygen and nutrient levels. The colouring it gives to water comes mostly from complex organic substances that have long been used to tan leather, and which are also abundant in some types of bark and leaves. Dark, peaty waters are usually relatively sterile compared to the teeming types of life that can be found even in ephemeral pools, partly because tannins are moderately toxic, but also because the tinted water reduces light penetration, so submerged plants can grow only in the shallowest waters around the fringes and phytoplankton is scarce.

In turn there is little in the way of zooplankton, so any larger animals present must feed mainly on any small fishes and sometimes tadpoles that can eke out a living among the emergent plants, or on insects and their larvae that scavenge for whatever washes into the deeper waters. Though these are rarely rich or complex habitats, the dark waters often look spectacularly beautiful, especially on sunlit days when the illuminated background stands out against the dark and reflective mirror below.

A black water river in south-western Australia.

Cormorants dry out in early morning sun after fishing in a cold lake.

So even though lakes may superficially look to be an ideal place for rich ecosystems to build up in their perennial waters, the combination of water that is too deep, oxygen-poor and cold to support most types of native fishes, little in the way of plants to attract herbivores, and with the few animals crawling around in the bottom sediments too far down for predatory birds to reach, some lakes may effectively be biological deserts. That isn't to say that some interesting creatures haven't evolved to meet the hidden challenges of deeper fresh waters despite the limited opportunities on this continent; at another extreme a shallow lake with winds, waves and currents constantly whipping up nutrients to fuel a plankton-fed explosion of life can be one of the most productive types of wetland, as we have seen in chapter 9.

Fishes and frogs

At higher altitudes fishes are scarce, yet there is a distinctive, as-yet unnamed species of mountain galaxias in Blue Lake just beneath the summit of Mount Kosciuszko, the highest freshwater lake in Australia and so pure that its waters are barely more saline than if the lake was filled with distilled water. Found in just two places with trout in all the waters between, the only reason this species hasn't disappeared from its last strongholds is the waterfalls along the connecting streams that have blocked the trout – for the moment.

With nothing much to feed on in the open waters of the lake, these fish seem to have lost the ability to recognise potential prey swimming above them, even mosquito wrigglers which are joyously devoured by most other galaxiids. Instead, they concentrate on scuds and other bottom-dwelling animals; this is probably why their relatively large eyes seem strikingly reflective in the lower half, an adaptation that possibly allows the upper half of the eye to scan the darker waters below with less distraction from the bright lights above.

The only native fishes that have specifically evolved to become freshwater lake dwellers are four other galaxiids that are found only in the central lakes of northern Tasmania and associated rivers. There were no really large predatory fishes here until the introduction of trout, yet these paragalaxias remain abundant in many places because they became bottom-dwellers a long time ago, lurking between rocks and in shallow vegetation to avoid predatory waterbirds; they can also be found in much deeper waters at times. Though the younger fish school in mid-depth water, adults act more like gobies than most galaxiids, perching slightly tilted on their front fins and skipping between and under rocks rather than swimming in the open.

The waters of the wallum country of northern New South Wales and southern Queensland seem to be a very different world at first glance – a mosaic of dark, peaty streams and shallow lagoons, merging abruptly in places with the upper

Mangrove ferns among sedges and paperbarks at the upper tidal limit of a wallum creek.

reaches of tidal estuaries, where soldier crabs may be found foraging over mudflats not far from where frogs are calling among rainforest ferns that fringe a wallum creek. Sedges and other emergent plants may form dense stands in places in the more tidal waters, but aquatic plants are scarce in these dark and peaty waters, and while aquatic animals are often also thinly spread, some seem to have a special affinity for these black waters.

Many of the aquatic animals visiting here are more widespread and don't have any particular ecological commitment to this type of water; this includes many fishes that breed in estuaries, while others such as purple-spotted gudgeon are also found much further inland. The honey blue-eye (once regarded as just another form of the variable Pacific blue-eye) is largely restricted to wallum waters, as is the Oxleyan pygmy perch. However, the stages of specialisation are most readily seen in the local rainbowfishes, including the crimson-spotted rainbow fish which is also found well to the north and south of the wallum belt in much clearer waters.

On sandy Fraser Island this species can be found in dune lakes that can be the colour of strong tea at times, and it seems to have made some headway in adapting to these specific conditions. When first caught, wild fishes from such locations are distinctly yellowish, with only traces of the blue and red seen in populations from clearer waters. Kept in clear water in ponds or aquaria they soon become less golden and, as the first captive-bred generation looks more like a washed-out version of the same species from clearer streams elsewhere, it seems that there must have been some degree of adaptation to dark, relatively lightless waters.

At another extreme the ornate rainbowfish is so distinctive that it forms a single-species group within this diverse family, and is so closely associated with these warm, peaty wetlands that it could almost be used to define the limits of the wallum. There seems to be a different colour or pattern variation of this species in most of the permanent coastal streams here, all of them evolved in isolation over the past ten thousand years as rising sea levels cut off many smaller streams from the larger rivers where their waters once merged.

For all of their obvious differences something that upland lakes, coastal streams and dune lagoons have in common is a marked absence of frogs. To some degree this may be caused by various combinations of low nutrient levels and sometimes acid waters, but it is also a response to the potential presence of predatory fishes. In higher and smaller water bodies that fishes haven't managed to reach, there is often at least one frog species present around the fringes, but in lower-lying areas including streams near the sea some very large and fast-moving fishes may move inland as the tide rises.

Hunting is obviously a part of their motivation as they race inland, and if frogs were breeding here their tadpoles would be easy prey, but it has also been suggested

Ornate rainbowfish.

that the tidal visitors are likely to be cleaning up external parasites, leaving only those that are able to survive the sudden transition from saline, alkaline estuaries to fresh and slightly acid waters.

Waterbirds and platypus: guides to hidden worlds

It isn't easy to see what lives in a large body of water where light doesn't penetrate far, but the presence (and also absence) of certain waterbirds is a reasonable guide to what goes on below the surface. Where the waters are reasonably clear cormorants will be among the most common visitors, but are only likely to stay for long periods of time if there is a reliable supply of smaller fishes. Once these have been depleted or become too nervous under hunting pressure, the cormorants are likely to come and go at unpredictable intervals, rarely staying more than a few hours as it doesn't take much to spook any surviving fishes that have already seen more than they want of their hunters.

The diving abilities of waterbirds are often confounded with those of sea mammals that can plunge to great depths, but with their hollow, air-filled bones there is a much shallower limit to how far any waterbird will find it worth diving. It takes hard work to get down to any depth for a buoyant animal, and also to stay there, so in deeper fresh waters where there isn't much in the way of

A platypus drifts along the shadowed shoreline of a small lake, just before retreating to its burrow as the day becomes brighter.

food to justify the effort even cormorants and grebes rarely bother going far under. And as these birds hunt by sight, there is no reason at all for them to stir up the mud or look for immobile prey in waters that are already too dark for them to see well.

That leaves potential food sources in slightly deeper, darker or murkier waters available for one of the most specialised mammals in the world, the platypus. In a small lake formed by a landslide within walking distance of where I live, a family of platypuses makes its living from the diverse insect larvae that feed on the organic matter constantly washing in from the surrounding forests and upstream. Closing their tiny eyes when they dive, platypuses must rely on other senses to find their often-miniscule prey and most of these seem to be built into their somewhat rubbery beaks.

Just how many senses are actually involved remains open to debate and should keep quite a few academics in work for many a decade to come, as it is almost impossible to study what is going on except in anaesthetised animals. The one thing that seems certain is that the beak is the essential hunting implement with its rich network of nerves and sensory equipment, which may include electroreceptors that can pick up minute electric signals from the nervous systems of their prey.

Great crested grebe.

Among all of the living waterbirds only grebes are even more specialised for an open water existence than cormorants, and even the stiffened fringe around their toes is likely to be a more efficient way to propel them under water than anything a cormorant can do with its more traditional webbed feet. The great crested grebe is a bird of lakes and deeper, more open swamplands, a creature so aquatic it is rarely seen out of the water, yet it is found across much of this planet and is one of the best-studied waterbirds due to its striking appearance and its complex and dramatic range of courtship behaviours.

Everything about this bird is exaggerated, and because its open-water ecological niche can easily become too crowded the males are among the most aggressive and territorial of all waterbirds, in extreme cases spearing competing males from below with their strong beaks or trying to drown them from the surface. Great crested grebes are smooth divers, swimming long distances submerged in pursuit of fishes and larger aquatic invertebrates. Even their young are good swimmers from an early age, looking remarkably like striped, furry and unsinkable toy turtles that would be very popular in the souvenir shop of any public aquarium.

Plant-eating birds are more attracted to the shallower and clearer waters of lakes, and aren't as fussy about the plant species they feed upon as about the water depth, depending on the length of their necks or their diving abilities. Dabbling ducks usually feed from the water's surface, bobbing around with only their attractively presented bottoms showing, or work their way along shorelines where they may undercut the edges with their foraging. Black swans have carried the process further, with their necks able to reach deeper beds of water plants comfortably from the surface; though it may seem an obscure point, I have sometimes wondered if the average length of the neck of any species of swan in any part of the world may be a concise average measure of the clarity of the waters it feeds in.

Eurasian coots are good divers which sometimes form large, tightly packed floating flocks, and I have watched a densely packed, circular raft of over a thousand of these foraging through widgeon grass in a slightly saline coastal lake for many days. Every hour or so a flicker of restlessness would start to ripple through the whole of the flock, until with a pounding beat of wings the entire group would lift off, only to land in the same circular formation 100 metres away. Although the flock never seemed to leave the water, every edible plant along the shoreline was trimmed to a nub, with only the tall-trunked swamp lilies standing, too high above their reach to be significantly damaged.

Eurasian coots.

17

Estuaries, mangroves and saltmarshes

The mixing zones where rivers run into the sea are called estuaries, and the water levels and salinity here are largely controlled by the rhythms of the tides. These often turbulent environments are among the most productive wetlands, nursery grounds to a multitude of fishes including commercially significant species as well as crabs and marine shrimps, so they are also a rich feeding ground for a wide range of carnivorous birds. Broad, slow rivers have extensive estuaries that may spread long distances inland, but the tidal zone of a small coastal creek that drops rapidly to the sea may be just a few metres long.

The most complex estuaries are the deltas of larger rivers, where sandbars created by waves may bank up the river waters so that a wide plain of fine sediments builds up behind, and floodwaters may cut new channels across the floodplain, as has already been described in chapter 16. If the tide sweeps in for long distances the finer sediments carried by the river will mostly end up out at sea, so the exposed flats at low tide will tend to be sandy rather than muddy; but if the higher tides spill out across low-lying ground they may create a mosaic of mangrove or saltmarsh to either side of the river.

A tidal floodplain gives a good idea of how billabongs form further inland, with the main channel flowing at left, the previous channel (no longer flowing) at right and in the foreground, and an old section of channel mostly filled with sand barely visible at centre.

Tea-coloured fresh water rides over the surface of waves as it rushes out to sea near Denmark in WA.

Plants: saltmarsh and mangroves

The dominant flowering plants around estuaries are low-growing saltmarsh species in the south, and diverse trees and shrubs known as mangroves further north. White mangrove is the most southerly species, a shrub in southern Victoria but much taller in warmer areas, where it is joined by an increasing range of mangrove species from many different plant families. These dense and impenetrable forests are a fine nursery for young fishes, crabs and other smaller animals and, in combination with the extensive paperbark swamps found in fresher waters upstream, help to reduce the erosional impacts of both floodwaters and extreme tides by slowing the rate at which they flow.

Aerial roots of the widespread white mangrove.

As the diversity of mangroves increases towards the tropics, a wider range of birds from groups more usually associated with terrestrial environments appears among them. In the far north these include specialist robins, whistlers and warblers as well as many honeyeaters, some of which visit only when food sources are abundant, while others are rarely found far from the mangrove stands. Even flying foxes often roost in large colonies among mangroves, not for their fruits but because the tangled roots and shifting silts below are a deterrent to potential predators.

The greater diversity of mangroves in warmer areas also means that the various species are spread across a wider range of habitats from seawater strength to nearly fresh waters. Although there may be extensive areas of tropical saltmarsh inland from the mangrove belt, there are few species in northern saltmarshes compared to the diverse herbaceous shrubs, grasses and sedges that abound in more temperate zones. In part this is because of the gentle slope of coastal plains in the far north, combined with the extremes of a hot dry season alternating with torrential flooding during the wet, which only a few plants are able to tolerate. Many saltmarshes are significant feeding grounds for various wetland birds as well as other animals, and even the endangered orange-bellied parrot migrates from Tasmania to mainland saltmarshes to forage through the summer months.

Zoning in a saltmarsh, separated from the sea beyond by a higher rise of grey saltbush.

Where tidal changes aren't dramatic, sea water may mix with fresh to create a saline gradient that decreases as it moves upstream, though this will migrate to and fro as the tide rises and falls. When the flow of fresh water is sluggish, a river estuary may become pure sea water at times, while on shorelines with strong wave action a sandbar may build up to cut all direct circulation between river and sea, flooding the ground behind the dunes with fresh water, within hearing range of the surf.

The highest (spring) tides that occur around the full and new moons can push a wedge of virtually undiluted sea water well upstream in a narrow river, even against a strong freshwater flow, so fluctuations in salinity can be dramatic. The most powerful tides may even create a kind of upstream tsunami known as a tidal bore in such confined conditions; though these can be impressive in parts of southern New Guinea, they are not particularly dramatic in Australia. At the other extreme, a flooding river which drains a very large catchment may ride over the denser sea water, remaining as a freshwater stream on the surface even some distance out from shore.

Shifting lives in estuaries

Nearly all of the animals that live in estuaries are well adapted to extremes, though some may move inland if salinity levels become uncomfortable for them, while others prefer to move out to sea. However, the more regular and powerful the water movement is, the more limited the range of plants that can grow in such situations, as these need to remain anchored in one place. There is little in the way of specialised estuarine plankton either, except in the largest and most sluggish estuaries, although marine species will often be washed in with the tide and create a food bounty for those creatures that don't mind a bit of turbulence.

White ibis.

Most of the larger animals that live in or visit estuaries can move about in pursuit of prey, and even the burrowing creatures can simply dig in more deeply during extreme tides or freshwater flooding. However, most of the food that is washed in by the tide, including such delicacies as decaying seaweeds and drowned animals, is easily accessible only when left exposed on sand or mudflats at low tide. Many of the burrowing animals, including scuds and slaters, some marine worms and other similar detritivores, would probably prefer to feed by night, but with only two low tides a day there isn't much choice for them, as feeding during higher waters would also expose them to schools of fishes patrolling the shallows.

When the invertebrates surface to feed, sometimes in great numbers, they draw shorebirds from stilts to oystercatchers, plovers, sandpipers and curlews, along with seagulls and other more generalised opportunists. The poorly named white ibis also often feeds across tidal flats, though it will happily forage in just about any other disturbed wetland, so not surprisingly it is often just a bit mud-stained.

Fishes of estuaries and streams

Schools of smaller marine fishes, such as hardyheads, follow the advancing tides, drawing more active hunters after them, so even the largest fishes prefer to move by night through the more open environments of estuaries. Cormorants are the perhaps the most familiar fish hunters in estuaries and bays throughout coastal Australia, though most of them are just as likely to be found well inland, where they adapt themselves to a diet of tadpoles, crayfish and shrimp as required. The black-faced cormorant is restricted in range to coastal southern Australia, while its distant relative the Australasian darter is most abundant in the north of the continent, although odd individuals can be seen be seen as far south as northern Tasmania. Its alternate name of snakebird is even more descriptive, not only for the long and slender neck, but also for the sinuous and reptilian way it sometimes uses its wings to manoeuvre underwater as it probes among rocks in clearer waters.

Egrets, herons and kingfishers will hunt fish anywhere from marine inlets to freshwater lakes, as long as water clarity is reasonable. Some tend to specialise in particular environments, particularly the eastern reef heron which roosts communally, with both grey and white individuals mixing together among mangroves or on islands. Even the relatively antisocial great crested grebe may gather in flocks in estuaries and enclosed bays outside its breeding season.

In southern Australia, the streams that feed estuaries often seem deserted during the day. Schools of common galaxias can be seen moving through rocky shallows, or in clearer waters a large goby may sometimes reveal itself by a brief movement. Most fishes and the abundant glass shrimps they feed upon lie low until dark, hidden from cormorants, ducks and the occasional kingfisher among

The small-mouthed hardyhead, a marine species that often ventures into fresh waters near the coast.

dense beds of reeds and rushes, or in grasses overhanging undercut banks. Sometimes the galaxias break into smaller foraging groups, writhing and probing among rounded pebbles where some smaller animal has taken refuge.

The common galaxias remains abundant despite the impacts introduced trout have made on many of its relatives, and its natural distribution makes it perhaps the most widely distributed freshwater fish in the world. Genetically similar populations are found from Western Australia to Tasmania, across the southern Pacific to Patagonia, and even to the Falkland Islands in the South Atlantic. The adults spend their lives in fresh waters: it is their tiny fry that make their way across oceans.

Another curiosity is that these galaxias are among the very few freshwater fishes anywhere in the world known to breed following the lunar cycle, the adults forming large schools around the time of the highest tides, mainly in autumn. How they recognise when the time is right is unknown; most of them migrate from well upstream where there isn't even a hint of tidal influence; it can't be the light of the moon either, as they may spawn around the full moon or the new moon. When the time is right and the spring tide pushes fresh water out over low-lying river banks, they lay their tiny eggs among grasses and other tussocks, to be left high but not too dry as the tide drops.

Most of the breeders are believed to die after mating, though some females may survive to return in later years; these larger fishes are more likely to be found in deeper pools upstream, and can look remarkably like a small and slender trout at first glance. Common galaxias always spawn upstream of the saltwater wedge that pushes its way inland on these very high tides, with hatching dependent on the

Common galaxias.

next spring tides 2 weeks later, though the eggs can survive up to 2 months if the tides aren't high enough to reach them until then.

There must be hundreds of billions of hatchling galaxias swept out to sea, returning as whitebait in the following spring. This one species dominates the returning swarms though there are also smaller numbers of other galaxiids, as well as Tasmanian whitebait south of mainland Australia. In New Zealand, where the common galaxias is known as inanga, the whitebait industry is estimated to catch half a billion of this single species annually, without significantly reducing the number of potential breeders. Young galaxias returning from the sea are transparent and empty-bellied, shrinking slightly and darkening as they adapt to fresh waters, as well as developing a different type of teeth and digestive system.

Their larger underwater predators come out only around dusk, with eels emerging from between rocks and other narrow places to probe their way along the shorelines. Tupong perch on the beds of leaves and snags where they were hidden during the day, or sit motionless on rocks around the deeper pools, watching for any small animal that may pass above their rotating, chameleon-like eyes. Spotted galaxias with their deep bodies and flared fins are also night hunters, and the larger ones are solitary and aggressive even to their own species, with a mouth capable of swallowing more slender galaxiids nearly a third of their length.

Their prey also moves in response to darkness, and the dense shoals of glass shrimp and small gobies that congregate among reeds and rushes spread out to feed, making it harder for the night hunters to find them. Even the schools of common galaxias which were so active during the day break up; unable to see each

Bullrouts may be abundant in some north-eastern rivers, but are rarely seen.

other in the dark, they scatter randomly through the shallower rocky rubble where their hunters are less likely to follow.

In warmer waters further north the extent of estuaries depends on the flow of the rivers which feed them. Rivers along the eastern Queensland coast are relatively short, dropping fairly quickly from the ranges, so their estuaries are often small in area compared to those of the big, sluggish rivers of the Top End. Along the eastern coast the superbly camouflaged bullrout is the warmer water equivalent of tupong

A pair of Pacific blue-eyes.

Origins: gobies (and gudgeons)

Gobies are the largest marine fish family, though many species are also found in estuaries and even further inland (see the desert goby in chapter 11). The gobies are mostly defined by their two pelvic fins, which are usually fused to form a more-or-less rounded suction disc, with a thickened and fleshy-looking base, that helps them cling to rocks in fast currents or disturbed waters. By contrast the closely related gudgeons have often been treated as a separate family in which the pelvic fins remain separate, but these simple definitions don't always hold true so the two families are often treated as one.

An impressive bicep – for a fish! A Takita's mudskipper emerges onto an algae-covered mudbank.

Mudskippers are among the most active, conspicuous and territorial gobies, and most of them are clearly gobies in the classical sense, yet a few odd species of this natural group such as Takita's mudskipper have separate pelvic fins even though they are in no other way like the more distantly related gudgeons. Many mudskippers live on tidal flats along the coast, where they fidget, jump and even somersault at low tide, their projecting eyes swivelling in all directions as they watch for potential food or threats. Others climb mangroves in more brackish waters, spending more time above water than in it, and may drown if kept submerged for too long.

By contrast, most other estuarine gobies are camouflage artists that are likely to be seen only if you hunt them by torchlight, or if you happen to be looking in the right place just when they make a sudden move. This family includes the smallest of all vertebrates, animals about the size and thickness of a little fingernail clipping, which are so frail and short-lived that they might almost pass as an invertebrate. Gudgeons are also common in many estuarine and inland waters, but these are mostly larger and many live in mid-depth water unlike the bottom-dwelling gobies.

A blue-spot goby props itself up on the fused pelvic disc which defines the true gobies.

in the south, rarely seen during daylight hours, but among the most common near-coastal predators, especially towards the tropics.

Despite its large mouth this species doesn't seem to be a particularly competent fish hunter, feeding mostly on the glass shrimps and long-armed prawns which breed prolifically in these warmer waters. While the tupong is no spinier than most other freshwater fishes, bullrouts are the only truly freshwater member of the scorpionfish family, with many venomous relatives including the stonefishes of the Indo-Pacific. Handle a bullrout carelessly, or brush against it with an unprotected foot while wading, and the intense pain caused by its spines may come and go for weeks, as the author can testify.

The tiny blue-eyes are closely related to the rainbowfishes, and though many species live their entire lives in fresh water, others are mostly estuarine. Even the Pacific blue-eye of the eastern coast, which can be found well inland at times, may retreat to the sea from fresher, cooler waters at colder times of year in its more southern range. Despite this tolerance of sea water Pacific blue-eyes don't seem to travel far, and some very distinctive local populations have developed through its wide range from central New South Wales to near Cape York.

Reptiles and fishes in the north

In the meandering rivers of the Top End and around the Gulf of Carpentaria the shallow gradients of the largest rivers, combined with an impressive tidal range, push murky, brackish waters well inland at high tide, so mangroves can be found a long way from the sea. In the wet season freshwater flows are predominant through the mangrove and saltmarsh zones; during the dry season it is the tides that push their way well inland instead.

Sharks, rays and sawfish may travel well inland along these broad waterways, and may even be found in pure fresh water for long periods of time, but the most conspicuous predator here is the saltwater crocodile. Reaching up to 6 metres in length, it can travel overland for considerable distances during the wet season, and as its numbers have increased over the past few decades a regular check must be made for ambitious strays even in the few officially safe swimming holes near Darwin.

The only other large, aquatic reptile found in the upper tidal reaches of the Top End is the pig-nosed turtle, the only non-marine turtle in Australia with flippers. Clumsy on land, it is difficult to see in turbid, moving waters, and although it was recorded from New Guinea well over a century ago it was recognised as a native only in the 1970s even though Aboriginal peoples had been both eating it and including it in rock art long before that.

Several species of aquatic snakes visit or live permanently in tropical mangrove areas, all of them strong swimmers that give birth to live young, so they don't have to leave the water to lay eggs. Some seasnakes enter estuaries, hunting over mudflats and among mangroves, and although all of these are extremely venomous few of them are aggressive, with the major exception of the estuarine seasnake. In South-East Asia this species is responsible for more than half of all seasnake fatalities, mostly through defending itself against fishermen when netted. Fatal bites from this species don't seem to be known in Australia, mainly because people are too afraid of the saltwater crocodiles which share these habitats to go near them!

Several monitor lizards (often called goannas in Australia) may be found among mangroves, though two are also widespread along watercourses further inland. The mangrove monitor is often seen resting in branches overhanging the water, dropping into the water if disturbed. This generalist predator and scavenger feeds on anything from insects to fishes and birds, and is a strong swimmer that readily moves out to sea, so it is also found around New Guinea and in many of the islands to its east.

Apart from the surface-feeding archerfishes and occasional schools of baitfish scattering at the strike of a predator from below, most estuarine fishes are hard to see in these turbid waters. Yet grunters, bream, diamondfish, scats and even young mangrove jacks, a well-regarded angling species, are common and widespread under the murk. Perhaps the most elusive and little-known estuarine fish is the nurseryfish, a high-headed, narrow-bodied species the male of which protects its eggs from suffocating in silt by dangling them from a peculiar hook on its forehead.

Mudskippers are common across much of tropical Australia, with some species at home on tidal mudflats along with the fiddler crabs that mark the closer presence of the sea, while others live and climb among mangrove roots further upstream. The place where the mudskippers become increasingly abundant, and the first fiddler crabs make their appearance as they work their way over the tidal flats, marks the beginning of the sea. Many other smaller, little-studied species of gobies are also found here, with new species still being named at times, and along the southern coasts these small but fast-breeding fishes are probably a significant food source for everything from larger fishes to piscivorous birds.

Other coastal wetlands

Not all coastal wetlands are tidal, and sea spray can drift considerable distances inland when large waves are breaking, turning even the purest lagoon waters slightly brackish. In wetlands that are constantly flushed by running fresh water, salt levels rarely build up much, so even the occasional seawater saturation during a storm is unlikely to have a long-lasting effect on the types of animals present. And

Young mangrove jack.

Fiddler crabs.

if there is any kind of barrier between the tidal zones and fresh waters, the transition from one to another can be abrupt and dramatic.

A sea-cliff waterfall at Cape Otway in southern Victoria is the most unusual example of a freshwater coastal wetland I'm familiar with; it is literally a vertical wetland which is washed by the sea at its base. The cliff itself is a rainbow of colours that give it its name: pink to orange algae, silvery limestone coated with a red bacterial mat, and even diverse dark-green mosses through which fresh water pours so fast that they aren't affected by seawater deluges during rough weather. Spectacular as they are even now, the falls have been three times as wide in wetter decades, as the broad faces of scoured silver, grey, orange and yellow limestone to either side testify.

There is no access for freshwater fishes here, making the cliff face and the springs directly above a specialised habitat for invertebrates. Scuds are common from just above high tide level at times, and various copepods and water fleas may appear in pools behind the top during dry periods when the freshwater flow is less. What lives in between will remain a mystery unless you are prepared to carry a very long ladder along the 3 kilometre, sandy, winding access track. Having carried a wetsuit and weight belt to the falls once, I'll leave this undoubtedly distinctive fauna for someone else to investigate!

To the south of the wallum country (see the previous chapter), the coastline of south-eastern Australia breaks into a great diversity of coastal wetlands ranging from drowned river valleys, such as Sydney Harbour with its numerous brackish inlets and arms, to large river deltas such as the Clarence, along with numerous coastal lakes linked with the sea in different ways, as far south as Mallacoota and the Gippsland Lakes in Victoria.

These lagoons, lakes and inlets are significant feeding grounds for young marine prawns, particularly the greentail, as far south as Mallacoota in far-eastern Victoria where the warm marine current that flows south along most of the east

Rainbow Falls.

coast moves out into the open sea. The cooler water temperatures in the estuaries further south support an increasingly different suite of species, a reminder that even the temperature of the sea itself has an effect on what can live in the wetlands it brushes up against.

Swamp lily flowering on the shores of the slightly saline Myall Lakes.

The Gippsland Lakes are a reminder of the other great force that now affects wetlands: human intervention. This very large system of interlocking lagoons in eastern Victoria was open to the sea only when floodwaters would break through the sandbars along Ninety Mile Beach, until a permanent opening was made at Lakes Entrance in 1889, to create better access for boats. Now salinity levels have risen in the areas closest to the entrance, and during the wetter times of year a surface layer of fresher water reduces oxygen interchange between the surface and the deeper areas.

As mangroves move further in towards areas where paperbarks were once the dominant plants, and increasing nutrient and sediment levels from cleared land surrounding the lakes create periodic blooms of algae and cyanobacteria, other submerged species disappear. The clearing of mangroves for unsustainable levels of coastal development also exposes acid sulfate soils to the air, compromising water quality still further: people who flock to build homes along the coast and rivers near the sea may be slowly destroying what they claim to love.

18

New opportunities: dams and other created wetlands

If you look out of the window of a plane flying over farmland on a cloud-free day, it is likely that at any one moment you will be able to count dozens of mostly regular-shaped silvery or blue pools below, some of them immense though many are so small they appear only as brief sparks as the sun reflects directly from them into your eyes. In agricultural areas, most of those glittering pools will be farm dams, and in many places they are virtually the only permanent waters in land that otherwise becomes parched during the dry season.

With their diverse sizes, shapes, depths and seasonal fluctuations these dams may harbour an impressive number of minute animals, and provide rest areas and feeding stations for migratory waterbirds and drinking water for other indigenous species. There is no census of the number of farm dams in Australia, and even if there were it would almost certainly overlook anything less than 50 square metres. It is possible that there could be a million or more of these varied and potentially useful habitats.

Many farm dams are so trampled and befouled by livestock that they will never support any real diversity of life, though it doesn't have to be so. It costs much less to fence livestock out of a dam than to rebuild or replace it as its walls erode, and silt builds up until it becomes so shallow that livestock must struggle to reach the water through a foul morass. And there are other, even more compelling economic reasons to reassess the primitive concept that dams are created only for cattle to bog themselves in, foremost of which are parasites and diseases.

An immature night heron fishes in the author's swimming dam near dusk.

If you know what the word 'hygiene' means, you already know there are very good reasons for keeping our drinking water and our waste products separate, yet livestock that are allowed free access to dams excrete straight into the water they drink from. And water (along with some of the smaller animals that live in it) is the medium through which diverse parasites and infections spread, from liver flukes to various diseases that affect both humans and animals.

An increasing number of studies, including in the USA, show that animals that are fenced out from dams so they can't keep re-infecting themselves and drink instead from troughs filled from the untainted water taken from that dam, grow much faster. Some sources suggest that livestock growth rates can increase up to 40% under these much more hygienic conditions, but even if you accept more conservative growth figures, healthier livestock could easily pay for the cost of fencing out the animals within just a year or two.

Once a dam has been fenced to keep livestock out, its water quality can be improved still further by turning it into a more attractive habitat, supporting a much wider range of species. If every one of Australia's million or so farm dams was improved for livestock in this way, they might also become a significant asset that helps to replace much of what we have drained or otherwise destroyed, instead of just providing a few more muddy waterholes. These and many related issues have been discussed in more detail in some of the author's earlier books (particularly *Wetland Habitats* and *Planting Wetlands and Dams*), but to convert ageing dams into potentially valuable habitats we have to be able to understand them through the eyes and other senses of the animals themselves.

What wetland animals need

Most humans have a special affinity with water, which is why real estate that looks out across wetlands, rivers and the sea is much more sought after and expensive than less well-endowed places. However, it is unlikely that most wetland animals have the same appreciation for an attractive waterscape as we do, so when thinking about what is most likely to attract them it should be in terms of food supply first, then shelter and, at certain critical times of the year, places to breed.

A heavily planted created wetland provides permanent habitat, including shelter for several species of waterbirds.

For animals that aren't necessarily able to move on easily from a dam, suitable breeding places are essential for their long-term survival. For example, freshwater turtles are long-lived animals that may give an impression of thriving if introduced to a dam, yet if there is no suitable place nearby for them to bury their eggs, or if these are likely to just be dug up and eaten by foxes they will just die out over time.

Most waterbirds live a more nomadic life, so a farm dam well stocked with their favourite foods and suitable places to perch or lurk can become a part of the mosaic of places which make up their lives.

If you are hoping to attract or introduce particular animals to a dam, your starting point should be to develop a broad knowledge of their natural history, general habits and preferred foods. In turn, that should give you a good idea of which types of local plants and also smaller animals are likely to be most useful for providing food or shelter. Older dams to which livestock have had free access for many decades may need to have excessive depths of silt removed as this can clog the gills of both fishes and crayfishes, and creating and planting shallow, seasonally flooded areas adjacent to the dam will increase the range of planktonic foods available after rain. Though dense plantings of trees and tall shrubs around a small dam are rarely desirable, a shelter belt that reduces noise from vehicles passing nearby will also drop snags that provide desirable habitat for both predators and their prey.

Attracting waterbirds

The success of a dam as a type of habitat is often measured by the diversity of waterbirds visiting, and this is as straightforward a measure as any others because it can also tell us about the diversity and abundance of other, less conspicuous

A family of wood ducks.

animals present. Some waterbirds such as ibises will even investigate a raw, newly filled dam, moving on within an hour or two if there is not much in the way of food, but others such as black ducks will nest in long pasture grass next to the tiniest dams as long as these have a substantial population of insects, snails and the like.

Common farm dam birds such as wood ducks may always live at the fringes of wetlands, retreating to the waters for protection from predators, but they spend much of their time grazing on grasses nearby and prefer tree hollows near the water for breeding. Black swans are mostly vegetarian and make their often immense, mounded nests from aquatic vegetation, though they don't seem worried about what the actual plants are as long as they are abundant and fairly soft. However, it is rare to see more than one breeding pair of swans in a dam, simply because they demand a certain amount of space between their nests.

Predatory species, including herons and cormorants, are unlikely to breed in a farm dam, but may include it on their regular feeding rounds as long as there is an adequate supply of tadpoles, small fishes or crayfishes to draw them back every few days. Young birds may be more adventurous while they still have only themselves to feed, and a young night heron spent several months lurking around our swimming dam until it began to develop adult plumage, at which stage it presumably moved on in search of a partner and more typical breeding habitat. For

A swan shakes off water as it returns to its nest, a mound of water-ribbon in a shallow farm dam.

predatory waterbirds, creating suitable habitats for their food animals is the key, including a range of shelter plants both above and below the water, and with a few snags or logs to perch and preen on.

Islands in dams are greatly overrated as an attractant, and it is rare to see them being used even as a resting place, let alone for breeding; the single most important factor in drawing a wide range of waterbird species to a dam or other wetland is probably its total surface area. Natural wetlands often cover many hectares of open ground, so that even hefty birds such as pelicans and swans that need a long run-up to take off or land can come and go easily. Landscaping with trees close to the water's edge has the opposite effect, because as these grow they increasingly reduce air movement and block flight paths, creating what must look like a deep, windless pit to passing waterbirds.

This effect was particularly obvious in a pair of very similar farm dams, less than 200 metres apart and separated only by a fence, that I was asked to comment on during a workshop. One was surrounded by trees already 6 metres tall and had no birds in or near it, while the other had been left unplanted and was being regularly used by small numbers of ibises and two species of ducks. A few islands large enough for trees scattered around an extensive and otherwise open wetland are not a problem, and with enough water between them and the shore will even be used as a roost by some species of waterbird. Widely spaced stands of smaller trees such as paperbarks standing in or beside the water may also be used as communal breeding sites by egrets, cormorants and ibises. However, even this can

An unusual use for a nestbox.

have its downside unless the wetland is extensive, as the concentrated droppings produced by a large colony may kill trees and adversely affect water quality in the vicinity.

Other inducements to breeding such as nesting boxes are often unattractive, and they are also often placed too much out in the open for waterbirds with a well-justified fear of swamp harriers. Though these have become accepted as a kind of low-level visual pollution in many urban wetlands, they are often left unused because none of the waterbird species they are intended for likes the wetland as an overall habitat. In a created wetland not far from where I live a dozen nest boxes that were placed more than a decade ago have been used only by swallows, though to be fair I have also seen them used to good effect by pelicans practising *tai chi*.

Fishes and frogs: finding a balance

Farm dams are often relatively isolated from their surrounding catchments, leaving them free of the various introduced fishes that have had dramatic impacts on many native fishes and frogs. Although frogs are likely to be able to make their own way to a dam, and will readily breed there if it has been appropriately planted, fishes must be deliberately introduced. Fishes from perennial streams or with a marine or estuarine stage in their life cycles, even those galaxiids that can adapt to land-locked lakes, are unlikely to ever breed in a dam, as dams rarely have suitable flowing waters, but there are many other, smaller species that could establish and multiply in still waters.

Many people who fence and plant dams for habitat purposes are often reluctant to introduce fishes of any kind, on the grounds that frogs can't establish in the same waters. While this may be true for garden ponds which are small, crowded and often short of living foods, frogs and fishes have coexisted for millions of years in many types of natural wetlands, and there is no reason they can't also do so in a farm dam with a little planning. Dense plantings around the shoreline provide breeding sites and help to protect tadpoles not only from fishes, but also from their many other predators, including cormorants, herons, kingfishers and egrets, though not from stealth predators such as mudeyes and water tigers, or even from backswimmers while the tadpoles are still small.

Even though the majority of tadpoles in a dam are likely to disappear down the throats of these diverse predators, most frogs have a secret weapon that prevents them from dying out: they breed prolifically. Each pair of breeders needs only two offspring to survive *on average* to maintain their population, and the size of that population will be strongly influenced by the resources available in a wetland or dam, not just by the number of predators. These resources can only be stretched so far, so a dam that can support (for example) a stable population of 500 adult frogs won't support ten or a hundred times as many; this is why some predators are

Pobblebonk frogs mating in the shallows of a well-planted dam.

needed to do the job of thinning the tadpoles out. In a stable ecosystem the great majority of tadpoles *must* be eaten, or the resulting excess will either starve or be forced to migrate, and most of the migrating animals are far more likely to be eaten than to reach a new home.

There are good reasons for not introducing large predatory fishes to dams which are intended to be useful habitats for a range of species, the most obvious of which is that a single large carnivorous fish has too great an appetite for a relatively small body of water. Yet there are many smaller fishes such as pygmy perch, hardyheads, smelt and gobies that may establish their own stable populations in a dam without bothering tadpoles much as long as there are other living foods available, while others eat tadpoles only while they are small or avoid certain types altogether. In southern Australia many frogs breed mainly during the colder months, and even in relatively small ponds three of my local frog species breed happily alongside Murray River rainbowfishes, purple-spotted gudgeons, and various carp gudgeons which don't start feeding until the water has warmed up, by which time the tadpoles are already far too large to be considered as prey.

Hidden worlds

Dams aren't the only potentially valuable wetland habitats on farmlands, and seasonally flooded places on even the most barren-looking paddocks may swarm at times with diverse but tiny animals. Unlike the larger animals such as birds, fishes and frogs, these short-lived miniatures often thrive on disturbance, so fencing is not needed to maintain their favoured environments. We have already seen one example of a shallow pool of this kind in chapter 7, where a combination of drowning pasture grasses and cattle manure fertilise algal blooms, in turn feeding a changing procession of microscopic animals over just a few months.

In deeper and longer-lasting ephemeral pools there is time for much larger animals to develop, the most intriguing example I know of being a shallow farm dam not far from where I live. This was probably originally just scraped out among rocks in a boggy patch of ground, drying out completely during drought years, but when it refills the diversity and abundance of life forms can be spectacular. Several species of water fleas and copepods appear within a month or two as you would expect, and though copepods dominate for the first few months the most abundant animal later is a form of helmeted water flea.

These are named for their high crests (see chapter 1), reminiscent of the Spanish helmets of the 17th century, and reputedly help protect them from

This unpromising looking farm dam supports diverse aquatic invertebrates, including relatively gigantic shield and fairy shrimps.

Fairy shrimps.

backswimmers. Even though these are among the largest water fleas, by the time their headgear is fully developed they have been outgrown by two of the largest planktonic crustaceans in this dam – shield shrimps and giant fairy shrimps, many of them as long as my little finger. In some years the fairy shrimps are all a uniform vivid scarlet, while in others they vary dramatically from bottle-green to pink, brown and rusty-red. Whether these colour differences have been caused by differing feeds, or separate batches of eggs from genetically different sub-populations hatching under different conditions, no one seems to know.

Backswimmers also become abundant as the smaller crustaceans proliferate in a dam, though waterboatmen are usually rare, and small beetles may appear in moderate numbers at times. And as the dam dries out through summer predatory waterbirds may begin to appear, in this case starting with a single yellow-billed

Yellow-billed spoonbills.

spoonbill. Although occasional individuals of this species can be seen in insect-rich farm dams for a few days, this one lingered for several weeks, and was joined at various times by ducks and even several terns that visited intermittently as the larger crustaceans become trapped in pools among the rocks.

To a casual observer there is no trace of the great changes taking place, and from the nearby highway all that the passing traffic observes is an irregular, rock-fringed, muddy dam, with an occasional avian visitor. The only way to follow the changes taking place below is to dip in a fine-meshed net every few weeks, though even that tells us little enough as the tiny players continue with their mysterious lives.

Across the whole of this continent there must be countless such dams and pools with diverse casts of players, probably including species no one has yet seen, let alone named. Even in farm dams there are hidden communities of animals that have not changed much in millions of years, and they will probably still look much the same even after the dams and any other traces of our present civilisation are long gone.

Created wetlands and habitat

Created wetlands have become a significant growth industry over several decades, but whether they do enough to justify the considerable amounts of money spent on them is a moot point. The one undisputed benefit of many created wetlands is that, if heavily planted, they are likely to improve water quality, especially of urban run-off. Despite the nebulous goal of 'creating habitat' which is invariably mentioned in plans for such projects, there is rarely any mention of what that habitat is for, or how it is to be managed. In reality, only a limited range of plants thrive in the raw, newly flooded terrestrial soils of a recently created wetland, and in turn these attract only a few of the most common and adaptable birds, most often black ducks; cormorants and ibises.

Waterbirds disturbed by one man and his dog passing by 100 metres away.

As created wetlands mature a somewhat greater range of waterbirds is likely to visit at times, but as these wetlands are mostly constructed in urban areas they are unlikely to become useful breeding environments unless they are extensive enough to minimise disturbance by dog walkers and children. There are some obvious exceptions such as the many water treatment wetlands at Werribee (near Melbourne), which are famous among birdwatchers for the range and numbers of waterbird species they attract. However, a large part of the reason such wetlands are so attractive to birds is that entry is by permit only, and pets, including dogs, are not welcome.

Another common problem with created wetlands is that they nearly all fill with introduced fishes (especially gambusia, goldfish and carp) as soon as water quality stabilises, making them valueless as habitat for most native fishes and frogs. Though you'd never guess it from the bewildered comments of the many people finding their newly created wetland swarming with aggressive, fast-breeding gambusia, the reason for these plagues has been clear for decades: most urban wetlands are built on low-lying ground, with no attempt made to prevent random invasions from upstream or during floods. On the positive side most predatory waterbirds are just as happy to eat introduced fishes as native species, and may move in to take advantage of the bounty as long as their other needs, including feeding and roosting areas well away from passing foot traffic, are catered for.

Glossary

aestivation: a state of dormancy during a hot or dry period.

amphipods: aquatic crustaceans, the ancestors of terrestrial leafhoppers which look much the same.

animal: in the broad zoological sense used in this book, any bird, invertebrate, fish, reptile or amphibian.

aquatic: living under water, whether fresh or marine.

artesian waters: underground waters brought up from some depth, used in drier areas as a source of stock and irrigation water, but usually very hard (see also hardness).

biomass: the collective weight of an animal or group of animals in a particular ecosystem.

biota: collective term for all living things including plants, animals, bacteria and fungi.

bloodworms: the bright-red larvae of chironomid flies.

brackish: a mix of fresh and saline waters, in no particular proportions.

catchment: the land surface from which the water to a wetland runs off.

copepod: tiny, free-swimming, shrimp-like crustaceans with no common name, often abundant and an important food for larger animals in both fresh and saline waters.

crustacean: a major invertebrate group with an exterior skeleton, distantly related to insects, and including shrimps, water fleas, copepods, crayfishes and leafhoppers.

cryptic: referring to unnamed populations of similar-looking animals, which are often found to be very different when their genetic make-up is looked at more closely.

cyanobacteria: blue-green 'algae'.

detritivores: animals that feed on organic debris, including the minute organisms on it.

endemic: found only in a particular area.

ephemeral: in wetlands, a body of water which regularly dries out for a part of the year.

estuary: the section near a river or creek mouth where sea and freshwaters mix (animals from such waters are *estuarine*).

Eurasian: originating on the combined Asian/European landmass.

exoskeleton: the relatively rigid skin of an arthropod, shed at intervals so it can grow.

exotic: introduced from another area, usually another country in the sense used in this book, but also a 'native' that doesn't belong.

fauna: animals of all kinds, from microscopic species to the more familiar birds, fishes and mammals.

filter-feeder: any animal that feeds on minute life forms by filtering them through some kind of mesh or web, usually specifically designed to capture specific prey sizes.

fry: newly hatched or very young fish.
genus: a group of related species, different from other genera (plural) in their family.
germination: the sprouting of seeds into plants.
groundwater: water pooled or flowing underground.
hardness (hard): a measure of the quantity of calcium, magnesium and other salts present in water, other than sodium salts (see also salinity).
herbivorous: feeding on plants.
hybridigen: a word coined by geneticists to describe much more complex relationships between species other than equal mixing of genes during mating.
hydrology: the study of water as it flows over- or under-ground, and associated changes.
indigenous: native to a particular area, in other words not introduced from elsewhere.
invertebrate: animal without a bony internal skeleton or backbone.
land-locked: an inland population (usually of fishes) normally needing a marine stage in their life cycles, and which have formed resident breeding populations in fresh waters.
larvae: immature stages of an insect, often very different in appearance from the adult stage.
limnology: the study of inland (most often fresh) waters and their biota.
lunette: crescent-shaped mound of sediments, usually built up on the fringe of a lake by wind, during a period of drought.
mandibles: insect jaws.
milt: fish sperm.
moult: shedding of an outer skin or of feathers.
mudeye: carnivorous larval stage of a dragonfly.
native: used loosely in this book to mean any animal or plant which originates in some part of Australia, and sometimes in the sense that it belongs where it is found (i.e. indigenous).
niche: a generalised term combining the ecological place and roles of any animal, including its relationships to other animals.
nymph: the immature, aquatic stage of many insects.
omnivorous: feeding upon both plant and animal foods.
organism: a general term for any living thing including bacteria, and possibly even including some viruses and related microscopic forms.
peat: waterlogged plant matter which has only partly decomposed because of the absence of nitrogen and some other nutrients, and which is very absorbent so that it acts as a water store in swampy soils.
pH: a scale used to compare acidity or alkalinity, centred around a neutral point of 7. pH readings below 7 indicate increased acidity as numbers go down, above 7 show increased alkalinity as the numbers go up.
photosynthetic: able to produce food directly from sunlight and simple, inorganic chemicals.
phytoplankton: photosynthetic plankton including diverse algae, but also less plant-like organisms.

piscivore: a carnivore specialising in fish.
plankton: freely drifting organisms, usually with little control over their direction of drift.
pupa: the intermediate stage between larval and adult stages in many insect groups.
raptors: hawks, falcons, ospreys and related predatory birds.
salinity: a measure of the quantity of sodium salts present in water (see also hardness).
soft: refers to water with low salinity and relatively few minerals.
spawning: fertilising eggs, in the case of aquatic animals often (but not always) outside the body in water.
spent: the term used to describe post-spawning fishes, which in many species die after spawning.
taxonomy: the study of the evolutionary relationships of animals and plants.
terrestrial: growing or living on land, rather than in water or very wet places.
twitchers: serious bird watchers, often in pursuit of new species to add to a life list of all species seen by them.
vector: an animal carrying or distributing the seeds, spores or other reproductive parts of various plants and animals, and also sometimes of diseases.
water table: the upper surface of underground water, which may rise or fall with the seasons.
zooplankton: animal plankton.

Further reading

Bennett A, Backhouse G & Clark T (1995) *People and Nature Conservation: Perspectives on Private Land Use and Endangered Species Recovery.* Transactions of the Royal Zoological Society of New South Wales, Sydney.

Brearly A (2005) *Ernest Hodgkin's Swanland: Estuaries and Coastal Lagoons of South-western Australia.* University of Western Australia Press, Perth.

De Dekker P (1986) What happened to the Australian aquatic biota 18 000 years ago? In *Limnology in Australia.* (Eds P De Dekker & WD Williams). CSIRO, Melbourne.

Lange K & Partners (1974) *A Preliminary Study of the Ecology and Hydrology of Lake Colac, Victoria.* The City and Shire of Colac, with Environmental Resources of Australia, Colac.

Lintermans M & Osborne W (2002) *Wet and Wild: A Field Guide to the Freshwater Animals of the Southern Tablelands and High Country of the ACT and NSW.* Environment ACT, Canberra.

McComb AJ & Lake PS (1990) *Australian Wetlands.* Angus and Robertson, Sydney.

Norris RH, Liston P, Davies N, Coysh J, Dyer F, Linke S, Prosser I & Young W (2001) *Snapshot of the Murray–Darling Basin River Condition.* Murray–Darling Basin Commission, Canberra.

Rogers K & Ralph TJ (Eds) (2011) *Floodplain Wetland Biota in the Murray–Darling Basin: Water and Habitat Requirements.* CSIRO PUBLISHING, Melbourne.

Romanowski N (2007) *Sustainable Freshwater Aquaculture: The Complete Guide from Backyard to Investor.* UNSW Press, Sydney.

Romanowski N (2010) *Wetland Habitats: A Practical Guide to Restoration and Management.* CSIRO PUBLISHING, Melbourne.

Saintilan N (Ed.) (2009) *Australian Saltmarsh Ecology.* CSIRO PUBLISHING, Melbourne.

Turner L, Tracey D, Tilden J & Dennison WC (2004) *Where River Meets Sea: Exploring Australias Estuaries.* Cooperative Research Centre for Coastal Zone, Estuary and Waterway Management, Brisbane.

Upfield AW (1983) *Death of a Lake.* Scribner Crime Classics. Scribner, New York.

Williams WD (1992) *The Biological Status of Lake Corangamite and Other Lakes in Western Victoria.* Report to the Department of Conservation and Environment, Colac. Adelaide.

Williams WD (Ed) (1998) *Wetlands in a Dry Land: Understanding for Management.* Environment Australia, Canberra.

Young W (Ed) (2001) *Rivers as Ecological Systems: The Murray–Darling Basin.*

Murray–Darling Basin Commission, Canberra.

Zeidler W & Ponder WF (Eds) (1989) *Natural History of Dalhousie Springs.* South Australian Museum, Adelaide.

Invertebrates

Readers with a particular interest in insects will find *The Waterbug Book: A Guide to the Freshwater Macroinvertebrates of Temperate Australia* (Gooderham J & Tsyrlin E (2002), CSIRO PUBLISHING, Melbourne) particularly useful; this also covers many other common wetland invertebrates in a general way. The only broad guide to a wider range of animals is *Australian Freshwater Life: The Invertebrates of Australian Inland Waters* (Williams WD (1980), Macmillan, Melbourne). This is widely available in libraries, though the names used are often dated so readers needing accurate names for some groups will need to refer to more recent and specialised works. There are as many detailed identification guides available as there are major groups of invertebrates, the widest array of which were published by the Murray–Darling Basin Commission (now the Murray–Darling Basin Authority) and can be sourced through its website at <http://www.mdba.gov.au>.

Anderson NM & Weir TA (2004) *Australian Water Bugs: Their Biology and Identification.* CSIRO PUBLISHING, Melbourne.

Boxshall GA & Halsey SH (2004) *An Introduction to Copepod Diversity.* The Ray Society, London.

Davis J & Christidis F (1997) *A Guide to Wetland Invertebrates of Southwestern Australia.* Western Australian Museum, Perth.

McCormack RB (2008) *The Freshwater Crayfish of New South Wales Australia.* Australian Aquatic Biological Pty Ltd, Karuah, NSW.

McCormack RB (2012) *A Guide to Australia's Spiny Freshwater Crayfish.* CSIRO PUBLISHING, Melbourne.

Patterson DJ (1996) *Free-Living Freshwater Protozoa: A Colour Guide.* UNSW Press, Sydney.

Ponder W & Lunney D (1999) *The Other 99%: The Conservation and Biodiversity of Invertebrates.* Transactions of the Royal Zoological Society of New South Wales, Sydney.

Suthers IM & Rissik D (Eds) (2009) *Plankton: A Guide to their Ecology and Monitoring for Water Quality.* CSIRO PUBLISHING, Melbourne.

Theischinger G & Hawking J (2006) *The Complete Field Guide to Dragonflies of Australia.* CSIRO PUBLISHING, Melbourne.

Fishes

Many regional books on fish species are dated in parts. Allen, Midgley & Allen (2002) is the most accessible recent source of (brief) updates on names and species, and includes all of the significant regional and national fish faunas in its recommended reading list. A more detailed summary of sources for south-eastern Australia can be found in McDowall (1996).

Allen GR, Midgley SH & Allen M (2002) *Field Guide to the Freshwater Fishes of Australia*. Western Australian Museum, Perth.

Larson HK, Foster R, Humphreys WF & Stevens MI (2013) A new species of the blind cave gudgeon *Milyeringa* from Barrow Island, Western Australia, with a redescription of *M. veritas* Whitley. *Zootaxa* **3616**(2): 135–150.

McDowall RM (Ed) (1996) *Freshwater Fishes of South-Eastern Australia*. Reed Books, Sydney.

Pusey B, Kennard M & Arthington A (2004) *Freshwater Fishes of North-Eastern Australia*. Centre for Riverine Landscapes, Griffith University, Nathan, Queensland.

Raadik TA (2011) Systematic Revision of the Mountain Galaxias *Galaxias olidus* Species Complex in Eastern Australia. PhD thesis. The University of Canberra.

Raadik TA & Kuiter RH (2002) Kosciuszko *Galaxias*: a story of confusion and imminent peril. *Fishes of Sahul* **16**(2): 830–834.

Romanowski N (2004) Notes on Dwarf Galaxias *Galaxiella pusilla*. *Fishes of Sahul* **18**(4): 80–86.

Schmidt DJ, Bond NR, Adams M & Hughes JM (2011) Cytonuclear evidence for hybridogenetic reproduction in natural populations of the Australian carp gudgeon (*Hypseleotris*: Eleotridae). *Molecular Ecology* **20**: 3367–3380.

Unmack PJ, Allen GR & Johnson JB (2013) Phylogeny and biogeography of rainbowfishes (Melanotaeniidae) from Australia and New Guinea. *Molecular Phylogenetics and Evolution* **67**: 15–27.

Wager R & Jackson P (1993) *The Action Plan for Australian Freshwater Fishes*. Australian Nature Conservation Agency, Canberra. This can be downloaded from <http://www.environment.gov.au/biodiversity/threatened/publications/action/fish>.

Useful websites

For information on the former distributions of most of the other larger fishes in our largest inland rivers, see <http://australianriverrestorationcentre.com.au/mdb/troutcod>.

The Australia New Guinea Fishes Association (ANGFA) publishes a quarterly journal, *Fishes of Sahul*, in high-quality colour, soon to be available online. Although this is primarily an aquarists association with a particular interest in rainbowfishes, the journal also regularly publishes articles by researchers on various other fish groups and sometimes on the larger invertebrates, and there are also local groups which meet regularly for talks and field trips. See <http://www.angfa.org.au> for contact details.

Reptiles and amphibians

Anstis M (2007) *Tadpoles of South-eastern Australia: A Guide with Keys*. Reed New Holland, Sydney.

Cann J (1998) *Australian Freshwater Turtles*. Beaumont Publishing, Singapore.
Cogger HG (2000) *Reptiles and Amphibians of Australia* (6th edition). Reed New Holland, Sydney.
Collins JP, Crump ML & Lovejoy TE (2008) *Extinction in Our Times: The Global Amphibian Decline*. Oxford University Press, New York.
Goldingay R & Osborne W (Eds) (2009) Ecology and conservation of Australian bell frogs. *Australian Zoologist* **34**(3): 325–460.
Heatwole H (2012) *Amphibian Biology Volume 10. Conservation and Decline of Amphibians: Ecological Aspects, Effect of Humans and Management*. Surrey Beatty & Sons, Chipping Norton.
Shine R (1998) *Australian Snakes: A Natural History*. Reed New Holland, Sydney.
Thompson MB (1983) Populations of the Murray River tortoise: the effect of egg predation by the red fox. *Australian Wildlife Resources* **10**: 363–371.
Tyler MJ (1999) *Australian Frogs: A Natural History*. Reed New Holland, Sydney.
Tyler MJ & Doughty P (2009) *Field Guide to Frogs of Western Australia*. Western Australian Museum, Perth.
Tyler MJ & Knight F (2009) *Field Guide to the Frogs of Australia*. CSIRO PUBLISHING, Melbourne.
Webb G & Manolis C (2002) *Australian Crocodiles: A Natural History* (2nd edition). Reed New Holland, Sydney.
Wilson S 2005. *A Field Guide to Reptiles of Queensland*. Reed New Holland, Sydney.

There are also numerous reptile and amphibian societies around Australia, loosely affiliated as the Australasian Affiliation of Herpetological Societies. A complete, fairly recent list of websites and addresses for these can be found in *What Snake is That? Introducing Australian Snakes* (Swan G & Wilson S (2008), Reed New Holland, Sydney), which also includes a good general summary of the biology and ecology of most of the more common snakes.

The site maintained by the Amphibian Research Centre at <http://www.frogs.org.au> includes contacts for local frog groups Australia-wide, and general regional guides to all Australian frogs with a useful selection of pattern variations, rather than just single photos to illustrate entire species.

Mammals

Barrett C (1946). *The Bunyip and Other Mythological Monsters and Legends*. Mail Newspapers, Adelaide.
Grant T (2007) *Platypus* (4th edition). CSIRO PUBLISHING, Melbourne.
Menkhorst P & Knight F (2001) *A Field Guide to the Mammals of Australia*. Oxford University Press, Melbourne.

Birds

Bird lovers are blessed with an abundant literature, usually well illustrated with photography and original artworks, as well as several identification guides which are included here without comment as to their relative merits. There are also many regional books which are potentially of much wider application and usefulness than their titles may suggest – to mention just one, *Birds of French Island Wetlands* (Quinn D & Lacey G (1999), Spectrum Publications, Melbourne).

Barker RD & Vestjens WJM (1989) *The Food of Australian Birds* (2 volumes). CSIRO PUBLISHING, Melbourne.

Frith HJ (1982) *Waterfowl in Australia.* Angus & Robertson, Sydney. Although long superseded by other and more specialised books, this remains an excellent introductory read for ducks, geese and swans.

Geering A, Agnew L & Harding S (2008) *Shorebirds of Australia.* CSIRO PUBLISHING, Melbourne.

Hollands D (1999) *Kingfishers and Kookaburras: Jewels of the Australian Bush.* Reed New Holland, Sydney.

Kingsford RT, Thomas RF & Wong PS (1997) *Significant Wetlands for Waterbirds in the Murray–Darling Basin.* National Parks and Wildlife Service, Sydney.

National Photographic Index of Australian Wildlife (1985). *The Waterbirds of Australia.* Angus & Robertson, Sydney.

Pizzey G & Knight F (2007) *Field Guide to the Birds of Australia* (8th edition). HarperCollins Publishers, Sydney.

Simpson K & Day N (2010) *Field Guide to the Birds of Australia* (8th edition). Viking, Melbourne.

Slater P, Slater P & Slater R (2003) *The Slater Field Guide to the Birds of Australia.* New Holland, Sydney.

The two major, nationwide bird conservation societies in Australia have recently merged to form Birdlife Australia, and if you have any interest in birds (not just wetland species) the moderate subscription fee includes the magazine *Australian Birdlife.* The news for some species that have been common until the last decade or two can be depressing, however. The new, combined website at <http://www.birdlife.org.au> is the best starting point for information on everything from groups with specialised interests, to tours, useful literature, contacts, and so on.

Plants

For information on weeds and their control, <http://www.weeds.org.au> is the single most useful site and includes links to many other sites, as well as online reporting of new arrivals or outbreaks, and current information on biological control.

Cowie ID, Short PS & Osterkamp Madsen M (2000) *Floodplain Flora: A Flora of the Coastal Floodplains of the Northern Territory, Australia.* Australian Biological Resources Study, Canberra.

Duke N (2006) *Australia's Mangroves: The Authoritative Guide to Australias Mangrove Plants.* University of Queensland, Brisbane.

Entwisle TJ, Sonneman JA & Lewis SH (1997) *Freshwater Algae in Australia: A Guide to Conspicuous Genera.* Sainty & Associates, Sydney.

Romanowski N (1998) *Aquatic and Wetland Plants: A Field Guide for Non-Tropical Australia.* UNSW Press, Sydney.

Romanowski N (2009) *Planting Wetlands and Dams: A Practical Guide to Wetland Design, Construction and Propagation* (2nd edition). CSIRO PUBLISHING, Melbourne.

Romanowski N (2011) *Wetland Weeds: Causes, Cures and Compromises.* CSIRO PUBLISHING, Melbourne.

Sainty GR & Jacobs SWL (2003) *Waterplants in Australia: A Field Guide* (4th edition). Sainty and Associates, Sydney.

Guide to common and scientific names used in this book

Albany pitcherplant – *Cephalotus follicularis*
Anchor worms – parasitic copepods of the genus *Lernaea*
Arafura file snake – *Acrochordus arafurae*
Archerfishes – several species of *Toxotes*
Australasian darter – *Anhinga novaehollandiae*
Australasian grebe – *Tachybaptus novaehollandiae*
Australian bass – *Macquaria novemaculeata*
Australian pelican – *Pelecanus conspicillatus*
Australian reed-warbler – *Acrocephalus stentoreus*
Australian shelduck – *Tadorna tadornoides*
Australian white ibis – *see* white ibis
Azolla – one of two native species of *Azolla*
Azure kingfisher – *Alcedo azurea*
Backswimmers – predatory bugs of the family Notonectidae
Baillon's crake – *Porzana pusilla*
Banjo frogs – several species of *Limnodynastes*
Banded stint – *Cladorhynchus leucocephalus*
Barcoo grunter – *Scortum barcoo*
Barramundi – *Lates calcarifer*
Barrow cave gudgeon – *Milyeringa justitia*
Barred frogs – *Mixophyes* species
Barred galaxias – *Galaxias fuscus*
Bell animalcules – *Vorticella* species, and some related genera
Bell frogs –*Litoria aurea* and *L. raniformis*
Berney's catfish – *Arius berneyi*
Bitterns – *Botaurus* and *Ixobrychus* species
Black catfish – *Neosilurus ater*
Black duck – *Anas superciliosa*
Black swan – *Cygnus atratus*
Black water worm – *Lumbriculus variegatus*
Black-banded rainbowfish – *Melanotaenia nigrans*
Black-faced cormorant – *Phalacrocorax fuscescens*
Black-headed skimmer – *Crocothemis nigrifrons*
Black-winged stilt – *Himantopus leucocephalus*
Blackfish – two species of *Gadopsis*
Bladderworts – *Utricularia* species
Blind cave eel – *Ophisternon candidum*
Bloodworms – the red larvae of chironomid flies
Blown grass – *Lachnagrostis filiformis*
Blue-eyes – fishes of the family Pseudomugilidae
Blue-spot goby – *Pseudogobius* species
Blue-spotted hawker – *Adversaeschna brevistyla*
Bony bream – *Nematalosa erebi*
Bream – several estuarine species of the family Sparidae
Brine shrimp – the introduced *Artemia salina* (there are also related native species)
Broad-leaved paperbark – *Melaleuca quinquenervia*
Broad-shelled turtle – *Chelodina expansa*

Brolga – *Grus rubicundus*
Brown treefrog – *Litoria ewingii*
Brown trout – *Salmo trutta*
Brownback crab – *Austrothelphusa transversa*
Buff-banded rail – *Gallirallus philippensis*
Bulgurru – the sedge *Eleocharis dulcis*
Bullrout – *Notesthes robusta*
Burrowing frogs – mostly *Heleioporus* and *Neobatrachus* species
Bushy starwort – *Aster subulatus*
Buttongrass – the sedge *Gymnoschoenus sphaerocephalus*
Cabomba – *Cabomba carolineana*
Cane toad – *Bufo marinus*
Carp – the introduced *Cyprinus carpio*
Carp gudgeons – various *Hypseleotris* species, including both inland species with complex hybrid relationships, and also some coastal species
Caspian tern – *Hydroprogne caspia*
Cattle egret – *Ardeola ibis*
Cat-tail – *Typha latifolia*
Cave gudgeon – *Milyeringa veritas*
Chequered rainbowfish – *Melanotania splendida inornata*
Cherabin – *Macrobrachium rosenbergii*
Chestnut teal – *Anas castanea*
Chironomid larvae – young of non-biting midges in the family Chironomidae
Chytrid (pronounced ky-trid) fungus – *Batrachochytrium dendrobatidis*
Cichlids – introduced fishes of the family Cichlidae (pronounced sick-lids)
Cisticolas – smaller relatives of reed-warblers, genus *Cisticola*
Climbing galaxias – *Galaxias brevipinnis*
Coal grunter – *Hephaestus carbo*
Cods – some fishes of the family Perchichthyidae
Comb-crested jacana – *Irediparra gallinacea*
Common blue-tail – *Ischnura heterostricta*
Common galaxias – *Galaxias maculatus*
Common reed – *Phragmites australis*
Common sideswimmer – various species of *Pseudomoera*
Common snakeneck turtle – *Chelodina longicollis*
Common swamp wallaby-grass – *Amphibromus nervosus*
Common water-ribbon – *Triglochin procera*
Common yabby – *Cherax destructor*
Congolli – *see* tupong
Copepod – tiny somewhat shrimp-like crustaceans; the species illustrated in chapter 1 is a *Boeckella*
Copperhead snakes – *Austrelaps* species
Coral ferns – *Gleichenia* species
Cordrushes – plants of the family Restionaceae
Cormorants – several species of *Phalacrocorax*, and others
Creeping water bugs – family Naucoridae; the species illustrated is *Naucoris congrex*
Crimson-spotted rainbowfish – *Melanotaenia duboulayi*
Cumbungi – either of the two native species of *Typha*, *T. orientalis* and *T. domingensis*
Curlew sandpiper – *Calidris ferruginea*
Cutting sedge – *Gahnia trifida*
Dabbling ducks – shallow-water species that feed from the surface, particularly *Anas* species
Dahl's treefrog – *Litoria dahlii*
Desert goby – *Chlamydogobius eremius*
Desert rainbowfish – *Melanotaenia splendid tatei*
Diamondfish – *Monodactylus argenteus*
Dragons – lizards of the family Agamidae

Dusky moorhen – *Gallinulla tenebrosa*
Dwarf flathead gudgeon – *Philypnodon macrostomus*
Dwarf galaxiids – several species of *Galaxiella*
Dytiscid beetles – a family with many groups of predatory waterbeetles; the larvae are also predators (see for example the *Chostonectes* larva in chapter 2)
Eacham rainbowfish – *Melanotaenia eachamensis*
Eastern great egret – *Ardea modesta*
Eastern little galaxias – *Galaxiella pusilla*
Eastern common froglet – *Crinia signifera*
Eastern dwarf treefrog – *Litoria fallax*
Eastern rainbowfish – *Melanotaenia splendida*
Eastern water dragon – *Physignathus lesueurii*
Eel-tailed catfishes –family Plotosidae, mostly species of *Neosilurus, Porochilus* and *Tandanus*
Egrets – mostly *Egretta* species
Estuarine crab – *Amarinus lacustris*
Estuarine seasnake – *Enhydrina schistosa*
Eurasian coot – *Fulica atra*
Fairy fern – *Azolla* species
Fairy shrimp – the species illustrated in this book are both *Branchinella* species, in chapter 1 *B. affinis* and the much larger *B. australiensis* in chapter 18
Fiddler crabs – some species of the family Ocypodidae
Fishing spiders – species of *Dolomedes*
Flathead galaxias – *Galaxias rostratus*
Flathead gudgeon – *Philypnodon grandiceps*
Flatworms – diverse species of planarians
Floodplain water-ribbon – *Triglochin dubia*
Flying foxes – several *Pteropus* species
Flyspecked hardyhead – *Craterocephalus stercusmuscarum*
Fork-tailed catfishes – family Ariidae, mainly *Arius* species
Freckled duck – *Stictonetta naevosa*
Freshwater catfish – *Tandanus tandanus*
Freshwater cobbler – *Tandanus bostocki*
Freshwater crab – *Austrothelphusa transversa*
Freshwater crayfish – all native species are in the family Parastacidae
Freshwater crocodile – *Crocodylus johnstoni*
Freshwater flathead – *see* tupong
Freshwater jellyfish – *Craspedacusta sowerbyi*
Freshwater mussel – many genera with different ecological needs; the species illustrated is a *Velesunio*
Freshwater prawns – various species of *Macrobrachium*
Freshwater shrimps – *Paratya* and *Caridina* species
Freshwater soles – several species of the family Soleidae
Froglets – *Crinia, Geocrinia* and *Paracrinia* species
Galaxiids – fishes of the family Galaxiidae
Gambusia – *Gambusia holbrooki*
Gebang palm – *Corypha utan*
Giant fairy shrimp – *Branchinella australiensis*
Giant glassfish – *Parambassis gulliveri*
Giant rush – *Juncus ingens*
Giant seed shrimp – *Australocypris* species
Giant water bugs – family Belostamatidae
Gill maggots – parasitic copepods of the genus *Ergasilus*
Glass shrimp – *Paratya australiensis*

Glassfishes – the family Ambassidae, mostly (but not all) species of *Ambassis*
Glassworm – larva of a phantom midge
Glassworts – two native species of *Sarcocornia*
Gobies – fishes of the family Gobiidae
Golden perch – *Macquaria ambigua*
Golden-headed cisticola – *Cisticola exilis*
Goldfish – *Carassius auratus*
Grasses – family Poaceae
Great cormorant – *Phalacrocorax carbo*
Great crested grebe – *Podiceps cristatus*
Great water beetle – *Onchohydrus scutellaris*
Green treefrog – *Litoria caerulea*
Green fishing spider – *Dolomedes* species (these also come in other colour forms including blue)
Green-and-gold grass frog – *Litoria aurea*
Greentail prawn – *Metapenaeus bennettae*
Grey saltbush – *Atriplex cinerea*
Grey teal – *Anas gracilis*
Growling grass frog – *Litoria raniformis*
Grunters – diverse fishes of the family Terapontidae
Gudgeons – fishes of the family Eleotridae
Gulf saratoga – *Scleropages jardinii*
Guppy – the introduced fish *Poecilia reticulata*
Hardhead duck – *Aythya australis*
Hardyheads – fishes of the family Atherinidae, mostly *Craterocephalus* species in fresh waters, but also *Atherinosoma* in estuaries
Helmeted water-fleas – *Daphnia carinata* and related species
Herons – mostly *Ardea* species
Hoary-headed grebe – *Poliocephalus poliocephalus*
Honey blue-eye – *Pseudomugil mellis*
Hydra – miniature, sea anemone-like species of *Hydra*
Inland paperbark – *Melaleuca parvistaminea*
Jabiru – *Ephippiorhynchus asiaticus*
Keelbacked water fleas – *Daphnia carinata* and relatives
Lampreys – fish-like animals in the families Geotriidae and Mordaciidae
Land crayfishes – various species of *Engaeus* in the south-east, *and Engaewa* in the south-west
Latham's snipe – *Gallinago hardwickii*
Lentil waterflea – *Pleuroxus* species
Leptocephalus – an old scientific name for the larval stage of various *Anguilla* eels and others, now used as a descriptive term for this leaf-like stage in an eel's life
Lignum (tangled) – *Muehlenbeckia florulenta*
Little file snake – *Acrochordus granulatus*
Long-armed prawns – *Macrobrachium* species, including both freshwater and marine species
Long-jawed spiders – *Tetragnatha* species
Long-necked turtles – various *Chelodina* species
Lotusbird – *Irediparra gallinacea*
Macquarie turtle – *Emydura macquarii*
Magpie goose – *Anseranas semipalmata*
Mangrove fern – *Acrostichum speciosum*
Mangrove jack – *Lutjanus argentimaculatus*
Mangrove monitor – *Varanus indicus*
Marbled eel – *Anguilla reinhardtii*
Marine clubrush – *Bolboschoenus caldwellii*
Marron – *Cherax tenuimanus*
Marsh frogs – various species of *Limnodynastes*

Marsh crayfishes – *Geocharax, Gramastacus* and *Tenuibranchiurus* species
Merten's water monitor – *Varanus mertensi*
Mexican waterlily – *Nymphaea mexicana*
Mosquitoes – blood-sucking flies of the family Culicidae
Mountain galaxias – *Galaxias olidus* and around 14 other related species soon to be named
Mountain shrimp – *Anaspides tasmaniae*
Mouth almighty – *Glossamia aprion*
Mudeye – a dragonfly larva, of any of dozens of genera
Mudskippers – *Periophthalmus* species; the photograph is of *P. takita*
Murray cod – *Maccullochella peelii*
Murray River spiny crayfish – *Euastacus armatus*
Murray River rainbowfish – *Melanotaenia fluviatilis*
Musk duck – *Biziura lobata*
Mussel-ostracod – *Mytilocypris* species
Nankeen night heron – *Nycticorax caledonicus*
Narrowleaf water-ribbon – *Triglochin lineare*
Native brine shrimp – *Parartemia* species; the photo is of *P. zietziana*
Needle bugs – several groups of elongated bug; the photo is of a *Ranatra* species
Nurseryfish – *Kurtus gulliveri*
Oblong turtle – *Chelodina oblonga*
Olive perchlet – *Ambassis agassizii*
Orange-bellied parrot – *Neophema chrysogaster*
Orange-headed waterbeetle – *Eretes australis*
Ornate rainbowfish – *Rhadinocentris ornatus*
Osprey (eastern) – *Pandion cristatus*
Oxleyan pygmy perch – *Nannoperca oxleyana*
Pacific azolla – *Azolla filiculoides*
Pacific black duck – *see* black duck
Pacific blue-eye – *Pseudomugil signifer*
Pacific heron – *see* white-necked heron
Painted snipe – *Rostratula australis*
Pandaka goby – *Pandaka lidwillii*
Paperbarks – diverse *Melaleuca* species
Paragalaxias – galaxiids of the genus *Paragalaxias*
Pea mussels – *Corbicula* species
Pearlfishes – annual *Cynolebias* species from South America
Pearl-ostracod – *Australocypris* species
Phantom midges – family Chaoboridae
Pig-nosed turtle – *Carettochelys insculpta*
Pink bladderwort – *Utricularia tenella*
Pink-eared duck – *Malacorhynchus membranaceus*
Plague minnow – *Gambusia holbrooki*
Platypus – *Ornithorhynchus anatinus*
Pobblebonk frog – *Limnodynastes dumerilii*
Prickly currant – *Coprosma quadrifida*
Protozoans – a major and diverse group of single-celled animals
Purple swamphen – *Porphyrio porphyrio*
Purple-spotted gudgeon – *Mogurnda adspersa*
Pygmy perches – *Nannoperca, Nannatherina* and *Edelia* species
Pygmy water-ribbon – *Triglochin alcockiae*
Quacking froglet – *Crinia georgiana*
Queensland lungfish – *Neoceratodus forsteri*
Rails – in wetlands, mainly *Rallus* species
Rainbow trout – *Oncorhynchus mykiss*
Rainbowfishes – fishes of the family Melanotaeniidae
Radjah shelduck – *Tadorna radjah*

Red saltpan algae – *Dunaliella salina*
Red-and-blue damselfly – *Xanthagrion erythroneurum*
Red-eared slider – *Trachemys scripta*
Red-kneed dotterel – *Erythrogonys cinctus*
Redclaw – *Cherax quadricarinatus*
Red-kneed dotterel – *Erythrogonys cinctus*
Red-necked avocet – *Recurvirostra novaehollandiae*
Redfin perch – *Perca fluviatilis*
River blackfish – *Gadopsis marmoratus*
River pandanus – *Pandanus aquaticus*
River redgum – *Eucalyptus camaldulensis*
Royal spoonbill – *Platalea regia*
Rushes – *Juncus* species
Rusty monitor – *Varanus semiremex*
Sacred lotus – *Nelumbo nucifera*
Salamanderfish – *Lepidogalaxias salamandroides*
Salmon catfishes – several *Arius* species
Salt-lake snail – *Coxiella* species
Saltwater crocodile – *Crocodylus porosus*
Saratogas – two *Scleropages* species
Sawsedges – numerous Gahnia species
Sawshell turtle – *Elseya dentata*
Scats – *Scatophagus* and *Selenotoca* species
Scorpionfishes – the mostly marine family Scorpaenidae
Seasnakes – family Hydrophiidae
Sedges – family Cyperaceae
Seven-spot archerfish – *Toxotes chatareus*
Sharp clubrush – *Schoenoplectus pungens*
Sharp-tailed sandpiper – *Calidris acuminata*
Shield shrimps – *Lepidurus apus* (illustrated) and *Triops australiensis*
Short-finned eel – *Anguilla australis*
Short-necked turtles – *Emydura* species
Shovellers – two species of *Anas*
Skinks – lizards of the family Scincidae
Silver gull – *Chroicocephalus novaehollandiae*
Silver perch – *Bidyanus bidyanus*
Slender marsh crayfish – *Geocharax gracilis*
Slipper animalcules – *Euglena* and related protozoans
Small-fruited water-mat – *Lepilaena bilocularis*
Small-mouthed hardyhead – *Atherinosoma microstoma*
Smooth crayfishes – diverse species of *Cherax*
Snapping turtles – *Elseya* species
South-western laceplant – *Aponogeton hexatepalus*
Southern freshwater prawn – *Macrobrachium australiense*
Southern frogs – diverse species of the family Myobatrachidae
Southern smelt – two fishes of the genus *Retropinna*
Southern pygmy perch – *Nannoperca australis*
Southern Victorian spiny crayfish – *Euastacus yarraensis*
Spangled perch – *Leiopotherapon unicolor*
Spanner-claw yabby – an unnamed *Cherax* species, sometimes confused with *C. rotundus*
Spiny crayfishes – various *Euastacus* species
Spoonbills – two species of *Platalea*
Spotted crake – *Porzana fluminea*
Spotted galaxias – *Galaxias truttaceus*
Springtails – the only aquatic genus in Australia is *Sminthurides*
Striped marsh frog – *Limnodynastes peronii*
Sundews – *Drosera* species
Swamp fern – *Stenochlaena palustris*

Swamp harrier – *Circus approximans*
Swamp lily – *Crinum pedunculatum*
Swamp paperbark – *Melaleuca rhaphiophylla*
Takita's mudskipper – *Periophthalmus takita*
Tall flatsedge – *Cyperus exaltatus*
Tall spikerush – *Eleocharis sphacelata*
Tasmanian giant crayfish – *Astacopsis gouldii*
Tasmanian mudfish – *Neochanna cleaveri*
Tasmanian whitebait – *Lovettia sealii*
Tassel cordrush – *Baloskion tetraphyllum*
Teals – several *Anas* species
Teatree – *Leptospermum* species
Threadfin rainbowfish – *Iriatherina werneri*
Tiger leech – *Richardsonianus* species
Tiger snake – *Notechis scutatus*
Tilapia – a generic name for a group of cichlid fishes including their hybrids, in Australia referring to *Oreochromis mossambicus*
Toadlets – *Pseudophryne* and *Uperoleia* species
Treefrogs – the true native treefrogs are mostly *Litoria* species, but this name is also sometimes applied to the entire family Hylidae
Trout cod – *Maccullochella macquariensis*
Tupong – *Pseudaphritis urvillii*
Variable groundsel – *Senecio pinnatifolius*
Varied sword-grass brown butterfly – *Tisiphone abeona*
Waterboatmen – bugs of the family Corixidae
Water button – *Cotula coronopifolia*
Water fleas – tiny crustaceans including species of *Daphnia*, *Moina*, etc.
Water hyacinth – *Eichhornia crassipes*
Water measurers – hydrometrid bugs
Water mites – the many and varied swimming species of order Acarina, and related to ticks
Water python – *Liasis mackloti*
Water rat – *Hydromys chrysogaster*
Water scorpions – bugs of the genus *Laccotrephes*
Water skinks – some *Eulamprus* species and forms
Water striders – bugs of the family Gerridae (the photo is a *Tenagogerris* species)
Water-ribbons – various *Triglochin* species
Waterlilies – *Nymphaea* species, including both native and introduced species
Weatherloach – *Misgurnus* species (in Australia probably *M. anguillicaudatus*)
Western rainbowfish – *Melanotaenia australis*
Western swamp turtle – *Pseudemydura umbrina*
Wheel animalcules – diverse species and groups of rotifers
Whirligig beetles – various species of the family Gyrinidae
Whiskered tern – *Chlidonias hybrida*
Whistling-ducks – two *Dendrocygna* species
White ibis – *Threskiornis molluca*
White mangrove – *Avicennia marina*
White-bellied sea-eagle – *Haliaeetus leucogaster*
White-faced heron – *Ardea novaehollandiae*
White-necked heron – *Ardea pacifica*
Widgeon grass – several species of *Ruppia*
Willow herbs – various species of *Epilobium*

Willows – diverse *Salix* species and hybrids
Wolf spiders – family Lycosidae
Wood duck – *Chenonetta jubata*
Yabby – *Cherax destructor*
Yellow-billed spoonbill – *Platalea flavipes*
Zebra duck – *Malacorhynchus membranaceus*

Index

Page numbers for images are shown in italic. See the separate guide to common and scientific names for specific photo identifications, and also some further comments on various species and groups.

Aboriginal peoples 39, 67, 197
acid sulfate soils 253
Albany pitcherplant 167, *167*
amphibians – differences from reptiles 61
Amphipoda – *see* scuds
animalcules 35
  bell 35, *36*
  slipper 36
  wheel 36–7, 98–100, 164
Anostraca – *see* fairy shrimps; brine shrimps
aquaculture 159
aquatic bugs – *see* water bugs
aquatic worms 39, 40, *41*, 221
Arafura file snake *210*
archerfish *56*, 56, 249
arthropods – definition 1
  exoskeleton and moulting 16
Australasian darter 155–6, *156*, 242
Australasian grebe 88, *105*, 142
Australian bass *51*
Australian pelican 89–91, *89–91*, 125, 134–5, 154, *260*
Australian reed-warbler *82*, 82, 132, 218
Australian shelduck 82, *83*, 142, 197
azolla *85*, 86, *206*
azure kingfisher *199*, 199

backswimmers *20*, 32, 111, 123, 136, 154, 219, 220, 264
backwaters – *see* billabongs
bacteria 98
Baillon's crake 85–6, *86*
banjo frogs 64–5
Barcoo grunter 184, *194*, 194
barramundi *147*, 153
barred frogs 187
barred galaxias 186
barriers – *see* waterfalls
Barrow cave gudgeon 175
beetles – *see* waterbeetles
Berney's catfish *52*
billabongs including formation 177, 203–4, *205*, *213*, *238*
bitterns 79, 213
black catfish *59*
black duck 97, *98*, 139, 259
black swan 80, 84, 125, 131–2, *132*, 133, *259*
black yabbies 102
black-faced cormorant 242
black-winged stilts *110*, 132, 139, *173*
bladderworts 167
blood and dissolved gases 3, 25
bloodworms *25*, 25, 39, 111, 135, 138
blown grass 132
blue-eyes 57, 174, 247–8
blue-spot goby *247*
blue-spotted hawker *23–4*
body segments fusing in arthropods 3
bony bream 118
brine shrimps 93, 145, 171–2, *172*
broad-leaved paperbark 113
brooks – *see* streams
brolga 79, 87, *109*, 110
brown treefrog including tadpoles *63*, 121–2, *122*, 123
brown trout *60*
brownback crab 14–15
buff-banded rail 141, *214*
bugs – *see* aquatic bugs

bulgurru 87, 110
bullrout *245*, 245–7
bunyip – reasons for disappearance 200–1
burrowing frogs 64
bush yabbies 102

cabomba 206
caddisflies including larvae 28, *29*, 221
cane toad 68, 69–70, *69*
capillary action 163
carbon dioxide – *see* gases
carnivorous plants 166–7, *166*, *167*
carp 60, *131*, 159
  and salinity 131
carp gudgeons 55, 262
Caspian tern 92, 140
cat-tail *206*
cattle egret *87*, 132, 139
cave gudgeon 175, *176*
chequered rainbowfish *211*
cherabin 14, *14*, 196
chestnut teal 83, *84*, 139, *140*, 142
chironomid flies
  adults – *see* non-biting midges
  larvae – *see* bloodworms
chytrid fungus – *see* frogs
Cladocera
  see water fleas
clam shrimps 9
climate change – potential effects 127–8
climbing galaxias 178
coal grunter *183*, 184
comb-crested jacana 212
common blue-tail damselfly *138*
common galaxias 134, 242–5, *244*
common reed 217–18
common snakeneck turtle 121, *204*
common water ribbon *105*
Conchostraca
  see clam shrimps
congolli – *see* tupong
copepods 4–7, *5*, *6*, *7*, *135*, 135–6, 142, 145, 172, 207, 263
  abundance 5
  and drought 5
  as a major food source 4–5
  in lakes 5
  major groups 5–6
cormorants 87–8, 125, 139, 155–6, 224, *229*, 233–4, 242
crab, freshwater 14–15, *16*, 101–2, *101*, 145
  estuarine 15
crakes 85–6, 141, 213
crayfishes 3, 13–14, 101, 145, 259
  migration 119
created wetlands *257*
  excessive populations of birds 260–1
  islands 260
  nest boxes 260
  trees – pros and cons 260
  see also Farm dams
creeks – *see* streams
crimson-spotted rainbowfish *57*, 192, *212*, 231–2
crocodiles 67, 156–7, *157*
crustaceans
  body plan and diversity 3–4
  ecological significance 4
  in ephemeral waters 97
  primitive species 145, 174
cutting sedge *104*
cycles – drought and flood
  alternating 130–1, 142, *154*, 172, 204, 211, 225, 263

dabbling ducks 83, 235
Dahl's treefrog 210
damplands 166
dams – *see* farm dams
damselflies 137, *138*, 151, 206, *221*, 221
  nymphs *15*, 19, 44, 130, 150–1, *150*, 220

Darling River 190
darter – *see* Australasian darter
Decapoda 12–13
desert goby 174, *174*
desert rainbowfish 119
diamondfish 120, *120*, 249
dotterels 92
dragonflies 21–2, 124, 137, 206–7, 221
  emergence 23–4, *23–4*
  flight 22
  larvae – *see* mudeyes
  mating 24
  naming 22–3
drainage 112, *112*
ducks 125, 213–14
dusky moorhen 81, *218*

eastern dwarf treefrog 70
eastern great egret *125*, 126, 139
eastern little galaxias 53–4, *209*, 209
eastern reef heron 242
eastern water dragon 187, *187*
ecology – broad definition 1
eel-tailed catfishes 58–9, *59*, 174, 183
eels 49–50, 178
  short-finned 49–50, *49*
  marbled 49
eggs of smaller wetland animals
  drought tolerance and survival 4, 8
  spread by wind and birds 4
egrets 87, 125, *128*, 139, 242
Eighteen Mile Swamp 215, *217*
environmental flows 78
ephemeral waters 97, 98, *99*, 103, 118
Ephemeroptera – *see* mayflies
estuaries 231, 237–42
  human impacts 253
  marine currents – influence on estuaries 251–2
  mixing of waters (and otherwise) *238*, 240–1
  movement of animals 241–2
  variations along Australian coasts 237, 245, 248, 251
Eurasian coot 138–9, 235, *236*
Ewens Ponds 168
exoskeleton explanation 1–2
eyeless cave animals 175–6, *176*

fairy shrimps 3–4, *4*, 145, 263, *264*, 264
farm dams 255–6, *263*
  as potential habitats 256–7
  isolation from natural wetlands as an advantage 261
  see also seasonal farmland habitats
feathers and moulting 76, 126, 197–8, 250
fiddler crabs 249, *250*
file snakes 69, 210–11, *210*
fishes
  adapting to inland waters 52–3, 116–19
  hunting strategies 153
  in created wetlands 261
  in ephemeral waters 107–9
  introduced 59–60
  limits to travelling abilities 47–8
  marine origins 48–9, 120–1
  mouth-brooding 55–6
  schooling 153
  senses 51–2, 58
fishing 159
flathead gudgeon *151*
flatworms 40–1
flies 25–7
flooding 128, 142, 171
floodplains 177, 203
flyspecked hardyheads *58*
food webs 148
fork-tailed catfishes *52*, 59, 183
foxes as egg predators 195
freckled duck 80, 224–5, *224*
freshwater basses and cods 53
freshwater crocodile 157, 183

freshwater jellyfish 41–2
freshwater lobster 182
froglets 64
frogs 47, 62–5, 105–7, 210, 222–3, 231, 232–3
  and chytrid fungus 122–3, 223
  fishes as predators 262
  in created wetlands including ponds 261
  in streams and rivers 187
  lifecycles 62–4
  migration 121–2
  tadpoles including as prey 62–3, *63*, 223, 224, 259, 261–2

galaxiids 120, 134, 167, *168*, 168, 178, 186, *186*, 209, 230, 261
  dwarf species 108, 209
  growth rates 134, 209
gambusia – *see* plague minnow
gas diffision 3
  and breathing 38–9
  see also gills
gebang palm *216*
genetic parasites 194
giant (freshwater) crayfish 182
giant glassfish 58
giant pearl-ostracod 171, *171*
giant rush *191*
gills and feeding 3
  physical *20*, 20
  tracheal 19
glass shrimp 14, 115, *116*, 196, 242, 244
glassfishes 58, *153*, 153, 183, 190
glassworms – *see* phantom midge larvae
goannas – *see* monitor lizards
gobies 54–5, 242, 244, 246–7, 249
  goby-like galaxiids 230
golden perch 53, 194
goldfish 159
Gondwanaland 72, 108, 186
grayling 185–6
Great Artesian Basin 174
great cormorant *76*
great crested grebe *234*, 235, 242
grebes 125, 234, 235
green treefrog 70, *73*
grey teal 139
growling grass frog 223–4, *223*
grunters 53, 183–4, *183*, 259
gudgeons 54–5, 174, 190
  Murray-Darling species 194
  relationship to gobies 246–7
  underground species 175–6, *176*

habitat needs – the animal perspective 257–8, 266
hardhead duck 126–7, *126*, 140, 198–9
hardyheads 57, *58*, 174, 242
heathlands 166
helmeted water fleas 9, 263–4
herons 87, 97, 125, 139, 154, 242
hoary-headed grebe 111, 141–2, *142*, 143
honey blue-eye 231
hunting of waterbirds 143, 159
hydra 42

ibises 86, 213, *214*, 259
inland paperbark 205
insects
  and flight 18–19, 29, 123
  body plan and structure 17–21, 149
  breathing and air 19–20, *20*
  compound eye 20–1
  in ephemeral waters 97, 219–21, *220*
  larvae in streams 179–80, *180*
  metamorphosis 26–7, *27*
invertebrates – definition 1
  breathing methods 221–2
irrigation 78–9

jade perch – *see* Barcoo grunter

kingfishers 88, 154, 242

lakes
 species diversity and abundance 227, 230
 freshwater 227, 230
 in sand dunes *228*, 228
 Lake Colac 129–43
 Lake Corangamite 129, 143–4, *144*, *170*
 Lake Eyre 95, 170
 saline 170–3
lampreys 50, 178
land crayfishes 13–14, *164*, 165–6, *165*
Latham's snipe 92
leeches 40, *42*
lentil water fleas 8
livestock effects 255–7, 258, 263
long-jawed spider 151, *152*
long-armed prawns – *see* prawns, freshwater
lotusbird – *see* comb-crested jacana

Macquarie turtle *66*
magpie goose 79, 82, *198*, 198
mammals 71–2
mangrove fern *231*
mangrove jack 249, *249*
mangrove monitor 248–9
mangroves 237, 239–40, 248
marine worms 242
marron *158*, 159, 208
marsh crayfishes 13, 102–3, *102*
marsh frogs 64–5
marshes – *see* swamps
mayflies including larvae 28, 179–80, *180*
Merten's water monitor *68*
Mexican waterlily *206*
midges – biting *and* non-biting 26
mites 42, 43, *43*
mogurndas 55
monitor lizards 68, *68*, 248–9
mosquitoes 25–6
 larvae 25–7, *26–7*, 147, 230
 pupae *27*
moss animals 41
moulting in arthropods 16, *16*
 see also feathers and moulting
mound springs 174
mountain galaxias 230
mountain shrimp 145
mudeyes 22, 148, 149–51, *150*, 207
mudskippers *246*, 247, 249
Murray cod 53, 191, *192*, 192–3
Murray crayfish 190, *195*, 196
Murray rainbowfish 119, 182, 262
Murray River *190*
Murray-Darling carp gudgeon *193*, 194
Murray-Darling river system 189, 190
 human impacts 200–1
 junction as a biological merging point 119, 189–90
musk duck 83, *84*, *146*
mussel-ostracod 9, *136*
mussels 37, 39–40, *197*, 197
 pea mussels *37*, 40

nankeen night heron – *see* night heron
narrowleaf water ribbon 103, *104*
needle bugs 32, *33*
night heron 87, *256*, 259
non-biting midges 136, *137*
Notostraca – *see* shield shrimps
nurseryfish 249

oblong turtle 121
Odonata – *see* dragonflies; *see also* damselflies
olive hymenachne 205
olive perchlet 58, *153*
orange-bellied parrot 240
orange-headed waterbeetle *124*
ornate rainbowfish 232, *232*

osprey 88
Ostracoda – *see* seed shrimps
oxbow lake – *see* billabong
Oxleyan pygmy perch 231
oxygen – *see* gases
oystercatchers 242

Pacific blue-eye 146, 231, *245*, 247–8
painted snipe 93
paperbarks 194
paragalaxias 186, 230
parasites 43–4, 256
  see also genetic parasites
pea mussels *37*, 136
pearl-ostracods 9
peat 166
  and effects on biological diversity 166–7, 228
  and water quality 168, 228–9
  see also wallum country
pelican – *see* Australian pelican
phantom midge larvae *19*, 26, 149
phytoplankton 6–7, 133, 135
Piccanninie Ponds *169*, 169–70
pig-nosed turtle 66, 248
pink bladderwort *166*
pink-eared duck 80, *110*, 111, 126, 140
plague minnow 59, *60*
plants
  aggressive interactions 216–17
  movement and migration 131, *133*
  in peaty waters 231
platypus *71*, 71–2, *233*, 234
Plecoptera – *see* stoneflies
plovers 92–3, 242
pobblebonk frog 65, *262*
prawns, freshwater 14, *14*, *196*, 196
predators
  hunting strategies 146, 153
  interactions with prey 147
protozoans 35–6
purple swamphen 132, 139, 225–6, *225*
purple-spotted gudgeon *55*, 231, 262
pygmy perch 53, 168, 185
pygmy water ribbon 103

Queensland lungfish 72, *72*

radjah shelduck 82–3
rails 85, 141, 213
rainbowfishes 56–7, *57*, 153, 183, 190, 209, 211–12, *211*, *212*
  moving south to colder waters 119
rays 248
red-and-blue damselfly *221*
red-eared sliders 70, *70*
red-kneed dotterel *77*, 140
red-necked avocet 93, *173*, 173
redclaw crayfish *12*, 119, 159, 208
redfin perch 184
reptiles
  differences from amphibians 61
rice farming and waterbirds 7
river blackfish 53, 167, *168*, 184–5, *185*
river red gum 191, *191*
rivers
  black-water *229*
  course changes 177
rotifers – *see* animalcules, wheel
royal spoonbill 86
rushes 217

sacred ibis – *see* white ibis
Sahul 116
salamanderfish *107*, 108–9, 209
salinity – *see* water
salmon catfishes 67
salt-lake snail *170*, 171
saltmarsh 237, 239–40, *240*
saltwater crocodile *62*, 157, 248
sandpipers 92, 242
saratogas 55–6
sawfish 248
saw-shelled turtle *67*

scorpionfishes 247
scuds 10–12, *11*, 242
seasnakes 248
seasonal farmland habitats 263–5
sedges 217
seed shrimps 3, 8–9, *10*, 100, *136*, 136
sharks 248
sharp-tailed sandpiper *77*, 92, 140, *141*
shelter – *see* snags
shield shrimps 10, *11*, 263, 264
shrimps 14, 145
  eyeless 175
sideswimmers – *see* scuds
silver gull 92, 242
silver perch *53*, 53
skinks 187
slaters 242
slender marsh crayfish *102*, 103
slipper animalcules 7
small-mouthed hardyhead *243*
smooth crayfishes 13, 196, 207–8, *207*
snags *148*, 148, 191, *229*
snails – *see* water snails
snakebird – *see* Australasian darter
snakes 68–9, 248
southern freshwater prawn *196*, 196
southern frogs 64
southern pygmy perch 53–4, *54*
southern smelt 57
southern Victorian spiny crayfish 181–2, *181*
spanner-claw yabby *207*, 208
spangled perch 116–17, *117*
sphagnum bogs 178
spiders 42, *45*, 151
spiny crayfishes 13, 168, 181–3, *181*
sponges 41
spoonbills 86–7, 139
spotted crake 141
spotted galaxias *134*, 134, 244
springs 163, 168–9, 174
springtails *100*, 100
stilts 83, 111, 173, 242
stoneflies including larvae 28, 179–80, *180*
streams
  and floodplains 177
  at higher altitudes 178, 178–80, 184
  flowing underground 167–8
  geographic variation 177
  seasonal 177, *178*
striped marsh frog *65*
swamp fern *216*
swamp harrier 87, 88, 261
swamp lily *252*
swamp paperbark *104*
swamps 215, *216*

tadpoles – *see* frogs
Takita's mudskipper *246*, 247
tall flatsedge *190*
tannins – *see* peat
Tasmanian mudfish 108
teal 83, *126*
temnocephalans *44*, 44
threadfin rainbowfish 211
tides and feeding in estuaries 241–2
tilapia 60
toadlets 64
treefrogs 64, 73
Trichoptera – *see* caddisflies
Troglodytes 175–6
trout 59–60, *60*, 159, 184, 186, 230
trout cod 193
tupong *120*, 120, 245
turtles, freshwater 65–7, 169, 195–6, 210, 257–8
  eggs and breeding 210
  hibernation and aestivation 66
  long-necked or snakeneck 66, 121, 196
  short-necked or snapping 66

variable groundsel 132
varied sword-grass brown butterfly 219, *219*
vertebrates – definition 1
Volvox 6

wallum country 230–2
water
  and erosion 95
  and salinity 163, 170–4
  chemical changes in contact with soil 161, 163
  hardness 168
  quality at high altitudes 163
  surface tension 221–2, *221*, *222*
water beetles
  dytiscids including larvae *18*, 29, *30*
  in caves 175
  larvae – *see* water tigers
water bugs 30–3, 152, 154
  creeping *31*, 32–3
  giant 32
water fleas 3, 7–9, *8*, *9*, 39, 136, 145, 207, 263–4
water rat 39, 71, 72, 73, 197
water ribbons 104–5, 259
water snails 37–8, 39, *39*, 219
water striders 30–1, *222*, 222
water tigers 146, 148, *151*, 151
waterbirds
  and parasites 43, 127
  as an indicator of habitat values 258–9
  eggs 80
  hunting strategies 154–6, 170, 233
  in billabongs 212–14
  in ephemeral waters 97, 109–10, 116
  migration 75–8, 79–80, 92–3, 126–7
  navigation 77–8
  see also feathers and moulting
waterboatmen *32*, 33, 111, 123, 136–7, 154, 264
waterfalls as barriers 47, 178, *179*, 250–1, *251*
waterlogged soils and animal life 164–5
weeds 205–6, *206*
western rainbowfish *119*, 119
western swamp turtle 66
wetlands – general definition 161
wheel animalcules – *see* animalcules
whirligigs 30, 222, *222*
whiskered tern 92, *92*, 140, *141*
whistling ducks 83, 126
white ibis 139, *241*, 242
white mangrove *239*
white-bellied sea-eagle 88
white-faced heron *98*, *148*
white-necked heron 81, 139, *139*
willow herbs 132
wood ducks 83, *258*, 259
worms 97, 13
  see also aquatic worms

yabby *146*, 159, 196, 208
yellow-billed spoonbill 264–5, *265*

zebra duck – *see* pink-eared duck
zooplankton 6–7, 133
  daily migrations and light 7, 111
  larval stages 12–13

www.ingramcontent.com/pod-product-compliance
Lightning Source LLC
LaVergne TN
LVHW060629110826
845147LV00014B/877

9780643107564